W0042912

Herausgeber:

Prof. Dr. *A. Davison*  Department of Chemistry, Massachusetts Institute
of Technology, Cambridge, MA 02139, USA

Prof. Dr. *M. J. S. Dewar*  Department of Chemistry, The University of Texas
Austin, TX 7812, USA

Prof. Dr. *K. Hafner*  Institut für Organische Chemie der TH
6100 Darmstadt, Schloßgartenstraße 2

Prof. Dr. *E. Heilbronner*  Physikalisch-Chemisches Institut der Universität
CH-4000 Basel, Klingelbergstraße 80

Prof. Dr. *U. Hofmann*  Institut für Anorganische Chemie der Universität
6900 Heidelberg 1, Tiergartenstraße

Prof. Dr. *K. Niedenzu*  University of Kentucky, College of Arts and Sciences
Department of Chemistry, Lexington, KY 40506, USA

Prof. Dr. *Kl. Schäfer*  Institut für Physikalische Chemie der Universität
6900 Heidelberg 1, Tiergartenstraße

Prof. Dr. *G. Wittig*  Institut für Organische Chemie der Universität
6900 Heidelberg 1, Tiergartenstraße

Schriftleitung:

Dipl.-Chem. *F. Boschke*  Springer-Verlag, 6900 Heidelberg 1, Postfach 1780

Springer-Verlag  6900 Heidelberg 1 · Postfach 1780
Telefon (06221) 49101 · Telex 04-61723
1000 Berlin 33 · Heidelberger Platz 3
Telefon (0311) 822001 · Telex 01-83319

Springer-Verlag  New York, NY 10010 · 175, Fifth Avenue
New York Inc.  Telefon 673-2660

**27** Fortschritte der Chemischen Forschung
Topics in Current Chemistry

# Nonaqueous Chemistry

Springer-Verlag
Berlin Heidelberg GmbH 1972

ISBN 978-3-540-05663-8     ISBN 978-3-540-36994-3 (eBook)
DOI 10.1007/978-3-540-36994-3

Das Werk ist urheberrechtlich geschützt. Die dadurch begründeten Rechte, insbesondere die der
Übersetzung, des Nachdruckes, der Entnahme von Abbildungen, der Funksendung, der Wiedergabe
auf photomechanischem oder ähnlichem Wege und der Speicherung in Datenverarbeitungsanlagen
bleiben, auch bei nur auszugsweiser Verwertung, vorbehalten. Bei Vervielfältigungen für gewerb-
liche Zwecke ist gemäß § 54 UrhG eine Vergütung an den Verlag zu zahlen, deren Höhe mit dem
Verlag zu vereinbaren ist.

© by Springer-Verlag Berlin Heidelberg 1972.
Ursprünglich erschienen bei Springer-Verlag Berlin Heidelberg New York 1972.

Library of Congress Cata- log Card Number 51-5497.

Die Wiedergabe von Gebrauchsnamen, Handelsnamen, Warenbezeichnungen usw. in diesem Werk
berechtigt auch ohne besondere Kennzeichnung nicht zu der Annahme, daß solche Namen im
Sinne der Warenzeichen- und Markenschutz-Gesetzgebung als frei zu betrachten wären und daher
von jedermann benutzt werden dürften

# Contents

# Conductance of Electrolytes in Organic Solvents

**Prof. Byron Kratochvil**

Department of Chemistry, University of Alberta, Edmonton, Alberta, Canada

**Prof. Howard L. Yeager**

Department of Chemistry, University of Calgary, Calgary, Alberta, Canada

## Contents

1

# I. Introduction

The use of nonaqueous solvents as media for studies of solvation and reaction mechanisms has become increasingly important in recent years. One technique that has clarified a number of aspects of ion behavior in nonaqueous solvents is that of conductance measurements. The renewed interest seen in this method has come about because of the development of expressions that successfully describe the dependence of conductance on concentration in dilute solutions. An additional factor contributing to its utility is that conductance measurements permit the estimation of ion pairing of slightly associated salts with good precision. Also, in conjunction with transport number measurements, single ion mobilities are obtained that give insight into the extent, and in some instances the nature, of ion-solvent interactions. Knowledge of ion mobilities is not only useful in the interpretation of solvent-solute behavior, but it also provides practical information for electrochemists working with electrolysis processes, in battery research, and so on.

The determination of accurate and precise limiting conductivities and ion association constants requires care in the design and use of the conductance apparatus, and in the purification and handling of solvents and salts. For this reason attention is given initially here to experimental aspects of conductance measurements. This is followed by a tabulation of selected data, primarily in dipolar aprotic solvents, and a brief discussion of data taken in one solvent, acetonitrile, which is intended to show the scope of interpretation possible at the present time.

# II. Experimental Aspects

The techniques and apparatus which have been developed to measure electrolytic conductivities in nonaqueous solutions have been adapted from aqueous conductivity measurements with some modifications. Direct current measurements suffer the limitation of requiring reversible electrodes — a serious limitation in nonaqueous solvents. Although this problem can be circumvented [1] in some instances, virtually all precision conductance data have been taken using the alternating current method. General descriptions of this method are given in several sources. [2,3]

## A. Apparatus and Materials

### 1. Electrical Components

A Wheatstone bridge, modified for use with alternating current, forms the basis of the measuring system. Usually two arms of the bridge are matched resistors of approximately 1,000 ohms, the third arm is the conductance cell and the fourth is a variable resistor and capacitor connected in parallel. Alternating current of low voltage is supplied by an audio frequency oscillator (adjustable to several frequencies) and the variable resistor and capacitor are adjusted until null balance is achieved. At null the value of the resistor is taken as the ohmic resistance of the solution. This procedure is repeated at several frequencies and the true solution resistance is obtained by extrapolation. The theory and design of alternating current bridges for electrolytic conductance measurements is discussed in Ref. [4-7].

The off-balance bridge signal is generally amplified, then detected by oscilloscope [8], cathode-ray "magic-eye" [9], or meter [10]. To avoid stray electromagnetic couplings the oscillator and detector are separated by a distance of several feet, all leads are shielded, and the shields grounded. For sharp null balance and to eliminate the effects of electrostatic couplings the detector connections to the bridge must be brought to exact ground potential. A "Wagner earthing device", which provides an adjustable impedance for sharpest null, is generally employed for this purpose. Finally, both oscillator and detector should be isolated from the bridge using transformers. Discussions of the precautions which must be taken in the design of the electrical apparatus are given in Ref. [2] and [3].

## 2. Conductance Cells

A well-designed conductance cell should possess the following properties:

1. The cell should give resistance values in the range of about 1,000–50,000 ohms for the desired concentration range [5].

2. A range of electrolyte concentrations should be made available in the cell without introducing impurities.

3. Stray capacitance-resistance paths from electrode leads and other sources should be minimized.

4. Effects from polarization of electrodes should be eliminated or properly treated to obtain the true ohmic resistance of the solution.

The relationship between *resistance and cell geometry* is given by:

$$\Lambda = \frac{10^3 \; \ell/A}{RC}$$

where $\Lambda$ is the equivalent conductance of a solution of concentration $C$ and resistance $R$. An approximate value of the cell constant is equal to the distance between the electrodes, $\ell$, divided by the cross-sectional area of the conducting solution. The cell should be designed to produce the optimum value of $\ell/A$.

A range of concentrations may be produced in three ways. Increments of salt [8] or concentrated stock solution [9,11] may be added to a given quantity of solvent, a concentrated solution may be successively diluted [11-13] or a series of solutions may be prepared independently [14]. The first two methods are more convenient than the third and are generally employed.

Several problems arise in the preparation of solutions in nonaqueous solvents. The large *thermal coefficient of expansion* of many solvents necessitates the use of weight methods to establish concentrations, with subsequent calculation of molarities from weight concentrations. Also, solutions must be prepared and maintained under strictly *anhydrous conditions* during the course of the experiment. Further, since the preparation of quantities of highly pure solvent is difficult, the use of minimum amounts is desirable. Finally, salts sometimes dissolve very slowly in certain solvents, which makes efficient stirring to hasten dissolution important.

Whether using dilution or concentration methods, care must be taken to exclude atmospheric water from the cell during all operations. Typically, all-glass systems are employed with an atmosphere of dry nitrogen or argon, and any transfers of salt or solution are performed in a glovebox. Care must be taken to use anhydrous salts of highest purity. If the salt is not analyzed before use, conductance measurements before and after successive purifications of the material should be compared as a check on purity [31].

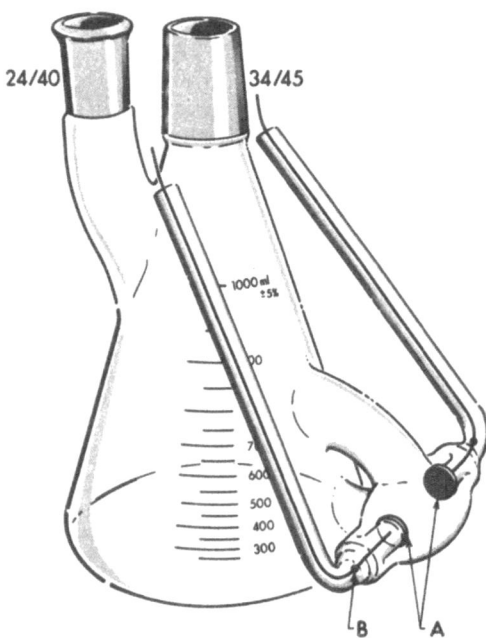

Fig. 1. Erlenmeyer conductance cell: 24/40 standard taper joint used for solvent delivery. *A* Platinum electrodes backed with glass for maximum rigidity. *B* Silver soldered platinum-copper junction

Kay and coworkers [8,15,16] use a technique that is simple, conserves solvent, and maintains anhydrous conditions. It employs an *Erlenmeyer-type cell* similar to that described by Daggett, Bair and Kraus [17] (Fig. 1) to which successive amounts of salt are added to a weighed portion of solvent by means of a *salt cup dispenser* (Fig. 2). The salt is weighed into Pyrex cups which are then loaded in the dispenser. When working with hygroscopic salts, capped salt cups are used and all transfers are performed in a glovebox. The solvent is delivered to the cell from a distillation reservoir through an all-glass system under $N_2$ pressure. The cell and dispenser are swept with argon gas while the dispenser is placed on the cell.

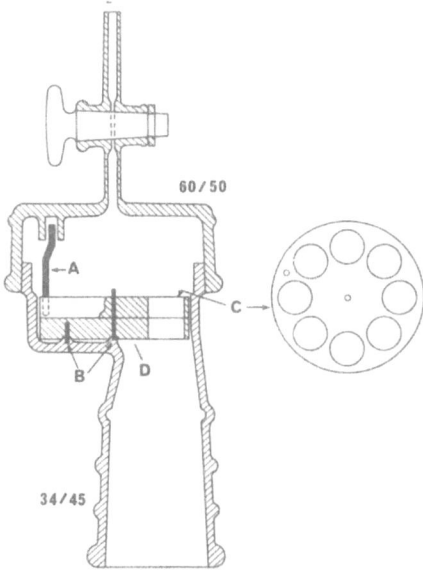

Fig. 2. Salt cup dispensing device: Metal pin *A* connects rotating Teflon disc *C* to upper glass joint; stationary Teflon disc *D* is anchored by pins at *B*

Another advantage of using this procedure is that the conductance of the solvent can be measured (and taken into account when calculating solution resistances) before any salt is added, with no new solvent added during the experiment. This Erlenmeyer-type cell has been used by several workers [9,10,12,18]. Various other designs include those of Jones and Bollinger [19], Shedlovsky [20], Nichol and Fuoss [21], Ives and Pryor [22], and Hawes and Kay [8]. These widely differing designs have been employed to overcome problems of stray capacitance-resistance paths and electrode polarization which lead to incorrect resistance measurements.

The Jones and Bollinger cell incorporates two circular disc electrodes in enlarged glass bulbs connected by a length of glass tubing. This design eliminates the *"Parker Effect"* [23], where the cell constant appears to vary with solution resistance. This effect disappears when cell filling tubes and electrode

5

leads are positioned so that stray currents are avoided. Dipping electrodes also suffer from the Parker effect and so cannot be used in precision work. Other sources of current leakage depend upon cell geometry and are indicated by a reduction in measured resistance with increasing signal frequency. At high resistance an opposite effect of increasing resistance with frequency has been observed; it has been attributed to a variety of causes [8,21,24,25].

Variations of resistance with frequency can also be caused by *electrode polarization*. A conductance cell can be represented in a simplified way as resistance and capacitance in series, the latter being the double layer capacitance at the electrodes. Only if this capacitance is sufficiently large will the measured resistance be independent of frequency. To accomplish this, electrodes are often covered with platinum black [2]. This is generally unsuitable in nonaqueous solvent studies because of possible catalysis of chemical reactions and because of adsorption problems encountered with dilute solutions required for useful data. The equivalent circuit for a conductance cell is also complicated by impedances due to faradaic processes and the geometric capacity of the cell [2,30]. In short, small but definite frequency dependence upon measured resistance is generally observed and the problem becomes one of obtaining the correct ohmic resistance of the solution by an extrapolation procedure. (See discussion of extrapolation procedures below.)

It is important to reduce this *frequency dependence* to a low level which can be extrapolated accurately. Nichol and Fuoss [21] have designed shielded dipping electrodes. An advantage of this approach is that electrode assemblies and thus the cell constant may be varied so as to incorporate a wide concentration range. A design by Ives and Pryor [22] incorporates two sets of electrodes in the same solution. The difference between observed resistances of the sets is measured, thus polarization effects are partly cancelled. Hawes and Kay [8] have constructed a cell in which one electrode is surrounded by a guard ring electrode to shield the inner electrode from the external environment. These designs involve formidable construction details, and although frequency dependence is reduced by their use, it is not eliminated completely.

The conventional method for determining the cell constant of a conductance cell involves the use of solutions of known specific resistance. The aqueous KCl solutions of Jones and Bradshaw [32] are the currently accepted standards. These workers carefully measured three solutions of given weight concentrations corresponding to molarities of about 1, 0.1 and 0.01. There are two disadvantages to this approach. First, a solution of an exactly specified concentration must be prepared. Second, it does not permit measurement of the cell constant over a range of concentrations in order to test for stray current leakages which would cause systematic variations in the calculated constant.

Fuoss and coworkers [33,34] solve these problems by calculating an averaged conductance equation for aqueous KCl solutions from the data of several workers. The equation reproduces the dilute Jones and Bradshaw standard value within experimental error, and has an estimated accuracy of 0.013 % up to a KCl concentration of 0.012 M. It has been found that cell constants measured over a wide concentration range vary as little as 0.02 % in Erlenmeyer-type cells [10, 15].

Solutions in a conductance cell are often stirred to hasten salt dissolution, to promote solution mixing, or to prevent temperature gradients. Some workers observe an upward drift in measured resistances of unstirred solutions [12,17] while others report a downward drift unless the unstirred solution is mixed by shaking of the cell immediately before the measurement [9,18,26]. The magnitude of this change is often 0.1 % or more. The effect has not been observed in other cases [8,10]. The source of this problem has been variously attributed to temperature variations, electrode adsorption effects and solvent impurities, although the problem has not been analyzed in detail. In all but one of the above cases [12] the resistance of the stirred solution was taken as the true value.

## 3. Temperature Control

The temperature coefficient of conductance is approximately 1 – 2 % per °C in aqueous [2] as well as nonaqueous solutions [27]. This is due mainly to the temperature coefficient of change in the solvent viscosity. Therefore temperature variations must be held well within ± 0.005 °C for precise data. In addition, the absolute temperature of the bath should be known to better than 0.01 °C by measurement with an accurate thermometer such as a calibrated platinum resistance thermometer. The thermostat bath medium should consist of a low dielectric constant material such as light paraffin oil. It has been shown [4] that errors of up to 0.5 % can be caused by use of water as a bath medium, probably because of capacitive leakage of current.

## 4. Solvent and Solute Purity

Accurate and meaningful conductance data may be obtained only in systems where the solvent and solute are free of foreign materials. Soluble conducting impurities in either one are obvious sources of error; less obvious are non-conducting impurities that effect solvation by competition with the solvent for coordination sites on ions. Purification of materials is always onerous, and is frequently aggravated by analytical difficulties in identification and measurement of trace contaminants.

Most nonaqueous organic solvents are purified by distillation, often at reduced pressure. Through careful fractionation the concentration of most impurities can be greatly reduced, though fairly large forecuts and tails may have to be rejected. Impurities that form azeotropes or have boiling points near that of the solvent can often be eliminated by conversion to more readily separated forms before the distillation step. Passage of the solvent through a column of molecular sieves to remove most of the water before distillation is often recommended. Fractional crystallization may be useful if the melting point of

7

the solvent is near room temperature and if decomposition tends to occur during distillation.

Criteria for solvent purity include careful measurement of physical constants such as melting point, refractive index, or conductance, but even these techniques may not detect some trace impurities. In many instances gas-liquid or other sensitive chromatographic methods may be used.

A number of solvents, particularily the amides, decompose on storage; decomposition is often catalyzed by ultraviolet radiation or by traces of impurities. In these cases it is important to use the solvent as soon after purification as possible.

The *purity of the electrolyte* is also critical in conductance work, and unfortunately much data otherwise carefully done is made suspect by the use of salts whose purity is assumed. In general the minimum purity of a salt for high precision work is 99.8 %, and 99.9 % or better is desirable. Wherever possible the purified salt should be analyzed by a reliable method having a precision of ± 0.1 %. If a method of suitable accuracy is not available the best alternative is to make a conductance run on a portion of the salt, then recrystallize the remainder one or more times from a suitable solvent and repeat the conductance run. If the conductance data for the two portions agree satisfactorily, the salt may be considered to be of acceptable purity.

An important source of error in nonaqueous conductance measurements is the presence of *water* in the system. As little as $1 \times 10^{-4}$ M water (2 mg/l) may cause errors in many solvents. The difficulties faced in maintaining anhydrous conditions are formidable. Closed cell systems for handling solvents and salts have been described earlier. The most widely used method for measuring the water content of a solvent at low levels is still the Karl Fischer titration.

## B. Treatment of Data

### 1. Frequency Extrapolation of Measured Resistances

The type and magnitude of frequency dependence upon measured resistance depends upon the design of the conductance cell. Generally, measured resistance decreases with increasing frequency, although the opposite effect is observed in some cases with Erlenmeyer-type cells [21,25]. Mysels *et al.* [25] analyze this effect and extrapolate to zero frequency on a plot of resistance *vs.* $f^{+2}$.

The more common procedure is to perform a linear extrapolation of resistance *vs.* $f^{-1}$ or $f^{-1/2}$, the value at infinite frequency being taken as the true resistance. For NaI solutions in acetonitrile it has been found that $f^{-1}$ plots yield straight lines at low concentrations, but $f^{-1/2}$ plots must be used to achieve linearity at higher concentrations [28]. Robinson and Stokes [2] discuss the causes of these variations.

Recently Hoover [29] compared various extrapolation methods for obtaining true solution resistances; concentrated aqueous salt solutions were used for the comparisons. Two Jones-type cells were employed, one with untreated electrodes and the other with palladium-blacked electrodes. The data were fitted to three theoretical and four empirical extrapolation functions by means of computer programs. It was found that the empirical equations yielded extrapolated resistances for cells with untreated electrodes which were 0.02 to 0.15 % lower than those for palladium-blacked electrodes. Equations based on Grahame's model of a conductance cell [30,7] produced values which agreed to within 0.01 %. It was proposed that a simplified equation based on this model be used for extrapolations. Similar studies of this kind are needed for dilute nonaqueous solutions.

## 2. Selection and Use of Conductance Equations

The equivalent conductivity of an electrolyte solution decreases with increasing concentration due to interionic attractions described mainly by the "electrophoretic" and "relaxation field" effects [2,35]. This decrease is more pronounced if in addition the electrolyte is associated. Association of ionic salts by ion-pairing is commonly observed in solvents of low or moderate dielectric constant. The immediate goals in the analysis of conductance data are the determination of the limiting equivalent conductance at infinite dilution, $\Lambda_0$, and the evaluation of the association constant, $K_A$, if ion-pairing occurs.

Various treatments of these effects have been developed over a period of years. The conductance equations of Fuoss and Onsager [36], based on a model of a sphere moving through a continuum, are widely used to interpret conductance data. Similar treatments [11,37], as well as more rigorous statistical mechanical approaches [38], will not be discussed here. For a comparison of these treatments see Ref. [11,38] and [39]. The Fuoss-Onsager equations are derived in Ref. [36], and subsequently modified slightly by Fuoss, Onsager and Skinner in Ref. [40]. The forms in which these equations are commonly expressed are

$$\Lambda = \Lambda_0 - SC^{1/2} + EC \log C + JC - F\Lambda_0 C$$

for unassociated electrolytes and

$$\Lambda = \Lambda_0 - S(C\gamma)^{1/2} + EC\gamma \log C\gamma + JC\gamma - F\Lambda_0 C\gamma - K_A C\gamma \Lambda f^2$$

where association is detected. Here $S$ is the Onsager limiting slope and the coefficients $S$, $E$, and $J$ are functions of $\Lambda_0$ and the dielectric constant, viscosity and temperature of the solvent. In addition $J$ is a function of an ion size parameter '$a$'. The $F\Lambda_0 C$ term corrects for viscosity changes; $F$ contains the hydrodynamic radius value $R$ [31,36]. The coefficient $F$ may also be set equal to the viscosity coefficient $B$ from the Jones-Dole viscosity equation [41,42]. For the small ions of typical inorganic salts this term becomes negligible.

9

In the second equation $K_A$, the association constant, is given by

$$K_A = \frac{(1-\gamma)}{C\gamma^2 f^2}$$

where $f$ is the mean ionic activity coefficient and $\gamma$ is the fraction of dissociated ions. For $K_A = 0$, $\gamma = 1$ and the second equation reduces to the first. The equations are valid for symmetrical univalent electrolytes up to a concentration corresponding to $\kappa a = 0.2$. The Debye-Hückel term $\kappa$ is given by

$$\kappa = \left(\frac{8\pi e^2 N \mu}{1000 D k T}\right)^{1/2}$$

where $\mu$ equals ionic strength and the other symbols have their usual meanings. Note that for lower dielectric constant solvents the upper concentration limit is correspondingly reduced.

Fuoss and Accascina [36] present graphical methods for treating conductance data according to either equation. Kay [8, 43] describes a computer program for least squares analysis in which standard deviations for the parameters are calculated as well. A similar program is described in Ref. [11].

## C. Determination of Single Ion Mobilities

The limiting equivalent conductance $\lambda_0$ is equal to the sum of cation and anion limiting conductances, $\lambda_0^+$ and $\lambda_0^-$. These quantities are related to the limiting transference numbers, $t_0^+$ and $t_0^-$, of the electrolyte by the equations

$$t_0^+ = \frac{\lambda_0^+}{\lambda_0^+ + \lambda_0^-} \quad ; \quad t_0^- = \frac{\lambda_0^-}{\lambda_0^+ + \lambda_0^-}$$

The experimental determination of transport numbers is discussed in Ref. [2, 44,45,46]. Due to serious experimental complications, highly accurate transference numbers have been measured in few nonaqueous solvents, with methanol [47], ethanol [48], nitromethane [49] and more recently acetonitrile [50] being the only exceptions. Transference numbers accurate to a few percent are available in most solvents, however.

Approximate single ion mobilities may be calculated by assuming that the cation and anion mobilities of a selected electrolyte are the same and equal to $\frac{1}{2}\Lambda_0$. Salts that have been used include tetrabutylammonium triphenylfluoroboride [51] and tetraphenylboride [52], triisobutylammonium tetraphenylboride [53], and tetraisoamylammonium tetraisoamylboride [54], the latter salt perhaps being the best choice. Kay and coworkers [50] have determined by high precision transference number measurements that the mobility of the cation of this electrolyte is 1.2 % smaller than the anion mobility in acetonitrile. This appears to be the limit of accuracy of this approach.

# III. Discussion of Conductance Data

## A. Tabulations of Conductance Parameters

### 1. Introduction

The principal factors affecting solvent-ion interactions can be classified as ion-dipole, Lewis acid-base, hydrogen-bonding, solvent structural, and steric. The solvent obviously plays a major part in these interactions. Therefore, to interpret trends in conductance data, bulk solvent properties such as viscosity and dielectric constant should be considered. Table 1 lists selected physical properties for a number of organic solvents.

### 2. Tabulations of Data

The fundamental information obtained from conductance data in nonaqueous solvents is the limiting equivalent conductance of a salt in a solvent and the degree of association between ions (that is, ion-pairing between the cation and the anion). Such information, coupled with transport number measurements, gives a way of obtaining single ion mobilities. This is the most important quantity, as it provides insight into the degree, and often into the nature, of solvent-ion interactions. Unambiguous conclusions are often difficult, however, as all of the effects noted above have to be considered. On the other hand, the technique of conductance is probably the most accurate one available at the present time for the determination of ion pair association constants of small magnitude.

Table 2 lists limiting equivalent conductance and association constant values for a number of 1 : 1 electrolytes in the solvents of Table 1, and Table 3 gives single ion mobility values. The data include results that appear to have sufficient precision to give meaningful values when treated by the Fuoss-Onsager conductance equation. In a few cases data of somewhat lower precision have been included to indicate the magnitude of the association constants, which can often be determined with fair accuracy from such data.

11

Table 1. *Selected physical properties of some organic solvents*[a]

| Solvent | Melting point (°C) | Boiling point (°C) | Density (g/cc) |
|---|---|---|---|
| Acetone | -95 | 56 | 0.791 (20 °C) |
| Acetonitrile | -45 | 82 | 0.777 |
| Adiponitrile | 1 | 295 | 0.958 |
| Benzonitrile | -14 | 191 | 1.001 |
| 1-Butanol | -90 | 118 | 0.806 |
| γ-Butyrolactone | -44 | 204 | 1.124 |
| *iso*-Butyronitrile | -72 | 104 | 0.765 |
| Dimethylacetamide | -20 | 166 | 0.937 |
| Dimethylformamide | -61 | 153 | 0.944 |
| Dimethylpropionamide | -45 | 175 | 0.921 |
| Methylethylketone | -87 | 80 | 0.805 (20 °C) |
| N-Methylformamide | -5 | 180 | 0.998 |
| N-Methyl-2-pyrrolidone | -16 | 202 | 1.028 |
| Nitrobenzene | 6 | 211 | 1.203 |
| Nitromethane | -29 | 101 | 1.131 |
| 1-Pentanol | -79 | 138 | 0.811 |
| 1-Propanol | -127 | 97 | 0.780 |
| Propylene carbonate | -40 | 242 | 1.206 (20 °C) |
| Pyridine | -42 | 115 | 0.987 (20 °C) |
| Sulfolane | 28 | 285 | 1.262 (30 °C) |
| Tetrahydrofuran | -108 | 66 | 0.888 (20 °C) |
| Valeronitrile | -96 | 141 | 0.795 |
| Dimethylsulfoxide | 18 | 189 | 1.096 |
| Ethanol | -114 | 78 | 0.785 |
| Ethylene carbonate | 36 | 246 | 1.322 (20 °C) |
| Formamide | 2 | 218 | 1.129 |
| Methanol | -98 | 65 | 0.792 |
| N-Methylacetamide | 30 | 206 | 0.942 (40 °C) |

[a] All values measured at 25 °C unless otherwise indicated.
[b] Data from Tables of Experimental Dipole Moments, A. L. McClellan, W. H. Freeman Co., San Francisco, 1963.
[c] Heats of reaction as a measure of the coordinating properties of a solvent toward a Lewis acid was suggested by V. Gutmann [148]. The values reported are negative enthalpies of coordination between the solvent and antimony (V) chloride, expressed in kcal per mole.

| Viscosity | Dielectric constant | Dipole moment[b] | $\Delta H_{SbCl_5}$[c] | Solvent classification[d] |
|---|---|---|---|---|
| (cp) | | (debyes) | kcal/ mole | |
| 0.32 | 20.7 | 3.0 | 17.0 | 6 |
| 0.35 | 36.0 | 3.37 | 14.1 | 5 |
| 5.99 | 32.4 | 3.7 | | 5 |
| 1.24 | 25.2 | 3.16 | 11.9 | 5 |
| 2.59 | 17.5 | 1.81 | | 2 |
| 1.75 | 39 (20 °C) | 4.03 | | 5 |
| 0.49 | 24.9 | 3.61 | 15.4 | 6 |
| 0.92 | 37.8 | 3.81 | 27.8 | 5 |
| 0.80 | 36.7 | 3.86 | 26.6 | 5 |
| 0.94 | 33.1 | | | 5 |
| 4.28 (20 °C) | 18.5 (20 °C) | 2.76 | | 6 |
| 1.65 | 171 | 3.82 | | 5 |
| 1.67 | 32.0 | 4.09 | | 5 |
| 1.98 | 34.8 (30 °C) | 4.0 | 4.4 | 7 |
| 0.62 | 35.8 (30 °C) | 3.1 | 2.7 | 7 |
| 3.48 | 15.0 | 1.66 | | 2 |
| 2.00 | 20.1 | 1.67 | | 2 |
| 2.52 | 65 | 4.98 | 15.1 | 5 |
| 0.97 | 12.3 | 2.3 | 33.1 | 6 |
| 10.30 | 43.3 (30 °C) | 4.81 | 14.8 | 5 |
| 0.55 (20 °C) | 7.6 | 1.75 | 20.0 | 6 |
| 0.69 | 20.0 | 4.12 | | 6 |
| 1.99 | 46.7 | 3.9 | 29.8 | 5 |
| 1.08 | 24.3 | 1.73 | | 1 |
| 2.55 | 81 | 4.87 | 16.4 | 5 |
| 3.30 | 109 | 3.4 | | 5 |
| 0.55 | 32.6 | 1.71 | | 1 |
| 3.02 (40 °C) | 165.5 (40 °C) | 4.23 | | 5 |

[d] The solvent classifications used here are: (1) solvents possessing both Lewis acid and Lewis base properties and a dielectric constant (D) $>$ 25; (2) solvents possessing both Lewis acid and Lewis base properties and D $<$ 25; (3) solvents possessing only Lewis acid properties and D $>$ 25; (4) same as (3) but D $<$ 25; (5) solvents possessing only Lewis base properties and D $>$ 25; (6) same as (5) but D $<$ 25; (7) solvents possessing negligible Lewis acid or base properties and D $>$ 25; and (8) same as (7) but D $<$ 25.

Table 2. *Limiting conductivities and association constants of selected 1 : 1*
All values at 25 °C unless otherwise indicated. Ph = phenyl, Pi = picrate, Octd =
Hept = heptyl, Ac = acetate, $SO_3Ph$ = phenyl sulfonic

| Salt | Acetone | | | Acetonitrile | | |
|------|---------|---------|------|--------------|-------|------|
| | $\Lambda_0$ | $K_A$ | Ref. | $\Lambda_0$ | $K_A$ | Ref. |
| HCl | | | | | | |
| HBr | | | | | | |
| HPi | | | | | | |
| LiCl | 214 | $3 \times 10^5$ | 58) | | | |
| $LiClO_3$ | | | | 170.0 | 400 | 59) |
| $LiClO_4$ | 187.3 | $5.3 \times 10^3$ | 60) | 173.0 | 4 | 61) |
| LiBr | 194 | $4.5 \times 10^3$ | 58) | | | |
| LiI | 195 | 145 | 58) | | | |
| LiPi | 157.7 | 819 | 63) | | | |
| $NaBPh_4$ | | | | 135.4 | 0 | 16) |
| $NaNO_3$ | | | | | | |
| NaSCN | | | | 189.8 | 87 | 64) |
| $NaO_3SPh$ | | | | | | |
| $NaClO_4$ | 191.2 | $4.3 \times 10^3$ | 60) | 180.4 | 10 | 16) |
| NaBr | | | | | | |
| $NaBrO_3$ | | | | | | |
| NaI | 183.6 | 177 | 67) | 179.4 | 0 | 68) |
| NaPi | 163.5 | 680 | 63) | | | |
| $KBPh_4$ | | | | 141.8 | 0 | 16) |
| $KNO_3$ | | | | | | |
| KSCN | 202.2 | $3.4 \times 10^3$ | 69) | 197.0 | 26 | 70) |
| $KClO_4$ | 188.7 | $7.1 \times 10^3$ | 60) | 187.6 | 17 | 71) |
| | | | | 187.5 | 14 | 16) |
| KBr | | | | | | |
| $KBrO_3$ | | | | | | |
| KI | 196.6 | 110 | 72) | | | |
| | 192.9 | 98 | 63) | 186.2 | 0 | 73) |
| | 197.5 | 179 | 58) | | | |
| KPi | 166.0 | 244 | 63) | | | |
| $RbBPh_4$ | | | | 143.8 | 0 | 16) |
| $RbClO_4$ | | | | 189.5 | 19 | 16) |
| $CsBPh_4$ | | | | 145.4 | 2 | 16) |

14

*electrolytes in several solvents*
n-octadecyl, Me = methyl, Et = ethyl, Pr = propyl, Bu = butyl, Am = amyl, Hex = hexyl,

| Benzonitrile | | | Dimethylacetamide | | | Dimethylformamide | | |
|---|---|---|---|---|---|---|---|---|
| $\Lambda_0$ | $K_A$ | Ref. | $\Lambda_0$ | $K_A$ | Ref. | $\Lambda_0$ | $K_A$ | Ref. |
| 1.5 | $4 \times 10^3$ | 55) | | | | 79.3 | $3.5 \times 10^3$ | 56) |
| | | | | | | 88.7 | 59 | 57) |
| | | | | | | 71.7 | 16 | 57) |
| | | | | | | 80.2 | $2.9 \times 10^2$ | 9) |
| | | | | | | 77.4 | 0 | 9) |
| 36.17 | $2.6 \times 10^3$ | 62) | | | | | | |
| 47.33 | 83 | 62) | | | | | | |
| | | | | | | 87.2 | 43 | 18) |
| | | | 74.6 | | 65) | 89.5 | 8 | 18) |
| | | | 18.4 | | 66) | | | |
| | | | 68.6 | | 65) | 82.2 | | 18) |
| | | | | | | 83.4 | 8 | 18) |
| | | | 21.8 | | 66) | | | |
| 48.11 | 53 | 62) | 67.6 | | 65) | 82.0 | 0 | 18) |
| | | | 57.2 | | 65) | 67.3 | 0 | 57) |
| | | | | | | 88.1 | 23 | 18) |
| | | | 74.2 | | 65) | 90.3 | | 18) |
| | | | 68.1 | | 65) | 82.8 | | 18) |
| | | | 68.4 | | 65) | 84.1 | 0 | 18) |
| | | | | | | 84.4 | 0 | 9) |
| | | | 22.0 | | 66) | | | |
| 52.12 | 77 | 62) | 67.0 | | 65) | 82.6 | 0 | 18) |
| | | | 56.8 | | 65) | 68.5 | 0 | 57) |
| | | | | | | 84.8 | | 9) |

Table 2 (continued)

| Salt | Acetone | | | Acetonitrile | | |
|---|---|---|---|---|---|---|
| | $\Lambda_0$ | $K_A$ | Ref. | $\Lambda_0$ | $K_A$ | Ref. |
| $CsClO_4$ | 199.9 | 223 | 74) | 191.0 | 22 | 16) |
| | | | | 191.2 | 23 | 71) |
| $NH_4ClO_4$ | | | | | | |
| $CuBF_4$ | | | | 173.1 | 9 | 10) |
| $CuPF_6$ | | | | 169.1 | 15 | 10) |
| $CuClO_4$ | | | | 168.4 | 0 | 10) |
| $AgBF_4$ | | | | 194.5 | 0 | 10) |
| $AgNO_3$ | | | | 192.4 | 70 | 10) |
| $AgPF_6$ | | | | 190.0 | 0 | 10) |
| $AgClO_4$ | | | | 189.7 | 0 | 10) |
| $TlBF_4$ | | | | 199.1 | 14 | 70) |
| $TlClO_4$ | | | | 195.2 | 32 | 70) |
| $Tl(CH_3)_2I$ | | | | | | |
| $Ph_4AsClO_4$ | | | | 159.6 | 0 | 50) |

| Salt | Dimethylsulfoxide | | | Ethanol | | |
|---|---|---|---|---|---|---|
| | $\Lambda_0$ | $K_A$ | Ref. | $\Lambda_0$ | $K_A$ | Ref. |
| HCl | 38.4 | 115 | 76) | 81.7 | 90 | 77) |
| $LiNO_3$ | | | | 42.7 | 19 | 78) |
| LiCl | | | | 38.94 | 27 | 43) |
| $LiClO_3$ | | | | | | |
| $LiClO_4$ | | | | | | |
| $NaBPh_4$ | 24.48 | | 81) | | | |
| NaAc | | | | | | |
| $NaNO_3$ | 40.8 | | 82) | | | |
| NaSCN | 43.0 | 0 | 82) | | | |
| $NaO_3SPh$ | 30.6 | | 82) | | | |
| NaCl | | | | 42.17 | 44 | 43) |
| $NaClO_4$ | 38.3 | 0 | 82) | | | |
| NaBr | 38.0 | 0 | 82) | | | |
| NaI | 37.6 | | 82) | | | |
| NaPi | 31.1 | | 82) | | | |

16

| Benzonitrile | | | Dimethylacetamide | | | Dimethylformamide | | |
|---|---|---|---|---|---|---|---|---|
| $\Lambda_0$ | $K_A$ | Ref. | $\Lambda_0$ | $K_A$ | Ref. | $\Lambda_0$ | $K_A$ | Ref. |
| | | | | | | 86.9 | 0 | [9] |
| | | | | | | 91.0 | | [9] |
| 52.18 | $2 \times 10^3$ | [62] | | | | 92.5 | 26 | [9] |
| 52.4 | | [75] | | | | | | |
| | | | | | | 87.6 | | [9] |
| | | | | | | 91.1 | | [9] |
| | | | | | | 79.3 | 13 | [9] |

| Ethylene carbonate (40 °C) | | | Formamide | | | Methanol | | |
|---|---|---|---|---|---|---|---|---|
| $\Lambda_0$ | $K_A$ | Ref. | $\Lambda_0$ | $K_A$ | Ref. | $\Lambda_0$ | $K_A$ | Ref. |
| | | | | | | 193.2 | 8 | [77] |
| | | | 25.5 | | [79] | 100.2 | 10 | [43] |
| | | | | | | 92.05 | 0 | [43] |
| | | | | | | 101.0 | 5 | [59] |
| 32.85 | | [80] | | | | 111 | | [9] |
| | | | | | | 81.76 | 0 | [134] |
| | | | 22.0 | | [79] | | | |
| | | | 27.5 | | [79] | 106.3 | 19 | [43] |
| | | | 20.5 | | [83] | | | |
| | | | | | | 97.4 | 0 | [43] |
| 38.84 | | [80] | | | | 116 | | [9] |
| | | | | | | 101.6 | 0 | [43] |
| | | | 26.7 | | [83] | | | |
| | | | | | | 92.05 | 0 | [84] |

Table 2 (continued)

| Salt | Dimethylsulfoxide | | | Ethanol | | |
|------|----------|-------|------|------------|-------|------|
| | $\Lambda_0$ | $K_A$ | Ref. | $\Lambda_0$ | $K_A$ | Ref. |
| $KBPh_4$ | | | | | | |
| $KNO_3$ | 41.5 | | 82) | | | |
| KSCN | 43.5 | | 82) | | | |
| KCl | | | | 45.42 | 95 | 43) |
| $KClO_4$ | 39.1 | 0 | 82) | | | |
| | 38.99 | | 81) | | | |
| KBr | 38.4 | 0 | 82) | | | |
| KI | 38.2 | | 82) | 48.2 | 50 | 85) |
| KPi | 31.7 | 0 | 82) | | | |
| $KOctdSO_4$ | 24.5 | | 82) | | | |
| $RbNO_3$ | | | | | | |
| RbCl | | | | | | |
| $RbClO_4$ | | | | | | |
| CsCl | | | | 48.33 | 158 | 8) |
| $CsClO_4$ | | | | | | |
| $NH_4NO_3$ | | | | | | |
| $NH_4Br$ | | | | | | |
| $AgNO_3$ | | | | 41.7 | 190 | 85) |
| | | | | 44.9 | 210 | 78) |
| $AgClO_4$ | | | | | | |
| $TlNO_3$ | | | | | | |
| TlAc | | | | | | |

| Salt | N-Methylacetamide (40 °C) | | Methylethylketone | | |
|------|----------|------|----------|-------|------|
| | $\Lambda_0$ | Ref. | $\Lambda_0$ | $K_A$ | Ref. |
| HCl | 20.6 | 66) | | | |
| HPi | 20.8 | 66) | | | |
| LiPi | | | 123.9 | $6.1 \times 10^3$ | 86) |
| $NaNO_3$ | 22.7 | 66) | | | |
| NaSCN | 24.2 | 66) | | | |
| NaCl | 17.8 (35 °C) | 87) | | | |
| $NaClO_4$ | 25.0 | 66) | | | |

| Ethylene carbonate (40 °C) | | | Formamide | | | Methanol | | |
|---|---|---|---|---|---|---|---|---|
| $\Lambda_0$ | $K_A$ | Ref. | $\Lambda_0$ | $K_A$ | Ref. | $\Lambda_0$ | $K_A$ | Ref. |
| | | | | | | 88.82 | 20 | 134) |
| | | | 30.1 | | 79) | 113.8 | 39 | 43) |
| | | | 29.9 | | 83) | 104.9 | 0 | 43) |
| | | | | | | 104.9 | 5 | 77) |
| 41.99 | | 80) | | | | 123.1 | 11 | 71) |
| | | | | | | 108.9 | 0 | 43) |
| | | | 29 | | 83) | 115.2 | 0 | 43) |
| | | | | | | 99.27 | 12 | 84) |
| | | | | | | 117.2 | 18 | 43) |
| | | | | | | 108.3 | 0 | 43) |
| 42.59 | | 80) | | | | | | |
| 43.59 | | 80) | | | | 113.4 | 15 | 43) |
| | | | | | | 131.5 | 33 | 71) |
| | | | 33.7 | | 79) | | | |
| | | | 31.7 | | 79) | | | |
| | | | | | | 110.8 | 60 | 77) |
| | | | | | | 121 | | 9) |
| | | | 33.2 | | 79) | | | |
| | | | 27.7 | | 79) | | | |

| N-methylformamide 88) | | Nitrobenzene | | | Nitromethane | |
|---|---|---|---|---|---|---|
| $\Lambda_0$ | Ref. | $\Lambda_0$ | $K_A$ | Ref. | $\Lambda_0$ | Ref. |
| 41.4 | | | | | | |
| | | | | | 122 | 89) |

19

Table 2 (continued)

| Salt | N-Methylacetamide (40 °C) | | Methylethylketone | | |
|------|------|------|------|------|------|
| | $\Lambda_0$ | ref. | $\Lambda_0$ | $K_A$ | Ref. |
| NaBr | 18.9 (35 °C) | 87) | | | |
| NaI | 20.6 (35 °C) | 87) | | | |
| NaPi | 20.2 | 66) | 125.7 | $2.2 \times 10^3$ | 86) |
| $KNO_3$ | 22.9 | 66) | | | |
| KSCN | 24.5 | 66) | | | |
| KCl | 17.9 (35 °C) | 87) | | | |
| $KClO_4$ | 25.2 | 66) | | | |
| KBr | 19.0 (35 °C) | 87) | | | |
| KI | 20.7 (35 °C) | 87) | | | |
| KPi | 20.2 | 66) | 131.7 | $0.8 \times 10^3$ | 86) |
| $KOctdSO_4$ | 15.4 | 66) | | | |
| CsBr | 20.0 (35 °C) | 87) | | | |
| $NH_4NO_3$ | 24.2 | 66) | | | |
| $NH_4ClO_4$ | 26.4 | 66) | | | |
| $NH_4Pi$ | | | | | |
| $AgClO_4$ | | | | | |
| $TlClO_4$ | | | | | |

| Salt | Propylene carbonate | | | Pyridine | | |
|------|------|------|------|------|------|------|
| | $\Lambda_0$ | $K_A$ | Ref. | $\Lambda_0$ | $K_A$ | Ref. |
| $LiNO_3$ | | | | | | |
| LiCl | 27.50 | | 94) | | | |
| $LiClO_4$ | 26.08 | | 94) | | | |
| LiBr | | | | | | |
| LiI | | | | | | |
| LiPi | | | | 58.6 | $1.2 \times 10^4$ | 96) |
| $NaBPh_4$ | | | | | | |
| NaSCN | | | | | | |

| N-methylformamide [88] | | Nitrobenzen | | | Nitromethane | |
|---|---|---|---|---|---|---|
| $\Lambda_0$ | Ref. | $\Lambda_0$ | $K_A$ | Ref. | $\Lambda_0$ | Ref. |
| 43.0 | | | | | | |
| 44.4 | | | | | | |
| | | 32.30 | $3.6 \times 10^4$ | 90) | | |
| | | | | | 130 | 89) |
| 41.9 | | | | | | |
| 43.7 | | | | | | |
| 45.0 | | | | | 122 | 89) |
| | | | | | 124 | 91) |
| | | 33.81 | $1.5 \times 10^3$ | 90) | | |
| 45.9 | | | | | | |
| | | | | | 128 | 89) |
| | | 34.4 | $7 \times 10^3$ | 90) | | |
| | | 38.4 | | 92) | 116 | 89) |
| | | | | | 124 | 89) |

| Sulfolane (30 °C) | | | Tetrahydrofuran [97] | | | Misc. | | |
|---|---|---|---|---|---|---|---|---|
| $\Lambda_0$ | $K_A$ | Ref. | $\Lambda_0$ | $K_A$ | Ref. | $\Lambda_0$ | $K_A$ | Ref. |
| 11.01 | | 93) | | | | | | |
| 13.63 | $1.39 \times 10^4$ | 95) | | | | | | |
| 11.05 | 6 | 95) | | | | | | |
| 13.25 | 278 | 95) | | | | 1-butanol | | |
| 11.53 | 6 | 95) | | | | 17.4 | 730 | 135) |
| | | | 88.5 | $1.2 \times 10^4$ | | N-Me-2-pyrrolidinone | | |
| | | | | | | 26.81 | 0 | 143) |
| | | | | | | adiponitrile | | |
| | | | | | | 9.16 | 0 | 142) |
| 13.20 | | 98) | | | | | | |

Table 2 (continued)

| Salt | Propylene carbonate | | | Pyridine | | |
|------|------|------|------|------|------|------|
| | $\Lambda_0$ | $K_A$ | Ref. | $\Lambda_0$ | $K_A$ | Ref. |
| NaClO$_4$ | | | | | | |
| NaI | 28 | 0 | 99) | 75.2 | $2.7 \times 10^3$ | 96) |
| NaPi | | | | 60.5 | $2.3 \times 10^4$ | 96) |
| KBPh$_4$ | | | | | | |
| KNO$_3$ | | | | | | |
| KPF$_6$ | | | | | | |
| KSCN | | | | | | |
| KCl | | | | | | |
| KClO$_4$ | 30.75 | | 101) | | | |
| KI | 31 | 2 | 99) | 80.4 | $4.8 \times 10^3$ | 96) |
| KPi | | | | 65.7 | $1.0 \times 10^4$ | 96) |
| RbSCN | | | | | | |
| RbClO$_4$ | | | | | | |
| CsBPh$_4$ | | | | | | |
| CsSCN | | | | | | |

| Sulfolane (30 °C) | | | Tetrahydrofuran [97] | | | Misc. | | |
|---|---|---|---|---|---|---|---|---|
| $\Lambda_0$ | $K_A$ | Ref. | $\Lambda_0$ | $K_A$ | Ref. | $\Lambda_0$ | $K_A$ | Ref. |
| 10.32 | 7 | 95) | | | | N-Me-2-pyrrolidinone | | |
| | | | | | | 41.78 | 0 | 143) |
| 10.29 | | 93) | | | | adiponitrile | | |
| | | | | | | 13.16 | 0 | 142) |
| | | | | | | dimethylpropionamide | | |
| | | | | | | 61.4 | | 9) |
| 10.86 | 5 | 95) | | | | l-propanol | | |
| | | | | | | 23.87 | 100 | 43) |
| | | . | | | | N-Me-2-pyrrolidinone | | |
| | | | | | | 41.62 | 0 | 143) |
| | | | | | | adiponitrile | | |
| | | | | | | 12.52 | 0 | 142) |
| | | | 90.1 | $3.1 \times 10^4$ | | | | |
| 10.74 | | 93) | | | | | | |
| 9.995 | 5 | 95) | | | | | | |
| 13.63 | | 98) | | | | adiponitrile | | |
| | | | | | | 15.91 | 0 | 142) |
| | | | | | | N-methylpropionamide | | |
| | | | | | | 10.22 | | 100) |
| 10.75 | 8 | 95) | | | | N-Me-2-pyrrolidinone | | |
| | | | | | | 41.95 | 0 | 143) |
| | | | | | | dimethylpropionamide | | |
| | | | | | | 60.9 | | 9) |
| 11.25 | 6 | 95) | | | | N-Me-2-pyrrolidinone | | |
| | | | | | | 41.49 | 0 | 143) |
| | | | | | | dimethylpropionamide | | |
| | | | | | | 58.8 | | 9) |
| | | | | | | adiponitrile | | |
| | | | | | | 13.26 | 0 | 142) |
| | | | | | | $\gamma$-butyrolactone | | |
| | | | | | | 46 | 0 | 99) |
| | | | | | | l-propanol | | |
| | | | | | | 25.69 | 230 | 43) |
| 13.87 | | 98) | | | | | | |
| 10.84 | 9 | 95) | | | | | | |
| 10.84 | | 93) | 108.7 | $5.2 \times 10^5$ | | | | |
| 13.96 | | 98) | | | | | | |

Table 2 (continued)

| Salt | Propylene carbonate | | | Pyridine | | |
|------|------|------|------|------|------|------|
| | $\Lambda_0$ | $K_A$ | Ref. | $\Lambda_0$ | $K_A$ | Ref. |
| $CsClO_4$ | | | | | | |
| $NH_4ClO_4$ | | | | | | |
| $NH_4I$ | | | | 95.2 | $4.2 \times 10^3$ | 96) |
| $NH_4Pi$ | | | | 80.5 | $3.6 \times 10^3$ | 96) |
| $AgNO_3$ | | | | 86.9 | $1.1 \times 10^3$ | 102) |
| $AgClO_4$ | | | | 81.9 | 520 | 102) |
| $AgPi$ | | | | 68.0 | 330 | 96) |

| Salt | Acetone | | | Acetonitrile | | |
|------|------|------|------|------|------|------|
| | $\Lambda_0$ | $K_A$ | Ref. | $\Lambda_0$ | $K_A$ | Ref. |
| $Me_4NBPh_4$ | | | | 152.3 | 0 | 103) |
| $Me_4NF$ | 182.7 | $1.1 \times 10^3$ | 63) | | | |
| $Me_4NPF_6$ | | | | 196.8 | 5 | 104) |
| $Me_4NCl$ | | | | 192.9 | 56 | 105) |
| $Me_4NClO_4$ | | | | 198.2 | 7 | 50) |
| $Me_4NBr$ | | | | 195.2 | 46 | 15) |
| | | | | 193.9 | 37 | 115) |
| $Me_4NI$ | 213 | 28 | 108) | 196.7 | 19 | 15) |
| $Me_4NPi$ | 183.4 | 67 | 109) | 171.8 | 1 | 15) |
| $Me_3PhNBr$ | | | | | | |
| $Me_3PhNI$ | | | | | | |
| $Me_3PhNO_3SPh$ | | | | | | |
| $Et_4NBBr_4$ | | | | 198.1 | 0 | 110) |
| $Et_4NBPh_4$ | | | | 142.8 | 0 | 103) |
| $Et_4NBPhCl_3$ | | | | 200 | 10 | 110) |
| $Et_4NNO_3$ | | | | 191.4 | 5 | 64) |
| $Et_4NCl$ | 194.2 | 370 | 67) | | | |
| $Et_4NClO_4$ | | | | 188.9 | 0 | 111) |

| Sulfolane (30 °C) | | | Tetrahydrofuran [97] | | | Misc. | | |
|---|---|---|---|---|---|---|---|---|
| $\Lambda_0$ | $K_A$ | Ref. | $\Lambda_0$ | $K_A$ | Ref. | $\Lambda_0$ | $K_A$ | Ref. |
| 11.03 | 9 | [95] | | | | | | |
| 10.95 | | [93] | | | | | | |
| 11.65 | | [93] | 11.65 | | | dimethylpropionamide | | |
| | | | | | | 62.7 | | [9] |
| | | | | | | l-butanol | | |
| | | | | | | 16.0 | 510 | [135] |

| Adiponitrile [142] | | | 1-Butanol | | | Dimethylacetamide [65] | | |
|---|---|---|---|---|---|---|---|---|
| $\Lambda_0$ | $K_A$ | Ref. | $\Lambda_0$ | $K_A$ | Ref. | $\Lambda_0$ | $K_A$ | Ref. |
| | | | 17.55 | $2.2 \times 10^3$ | [106] | | | |
| | | | 17.88 | $2.1 \times 10^3$ | [106] | | | |
| 12.93 | 28 | | | | | 70.1 | | |
| 13.27 | 0 | | | | | | | |
| 11.80 | 14 | | | | | | | |

Table 2 (continued)

| Salt | Acetone | | | Acetonitrile | | |
|------|---------|---|------|--------------|---|------|
| | $\Lambda_0$ | $K_A$ | Ref. | $\Lambda_0$ | $K_A$ | Ref. |
| $Et_4NBr$ | | | | | | |
| $Et_4NI$ | 207 | 60 | 108) | 187.3 | 5 | 112) |
| $Et_4NPi$ | 176.6 | 45 | 63) | 164.6 | 10 | 113) |
| $(EtOH)_4NBPh_4$ | | | | 122.3 | 0 | 87) |
| $(EtOH)_4NI$ | | | | 166.0 | 143 | 87) |
| $Pr_4NBPh_4$ | | | | 128.4 | 0 | 103) |
| $Pr_4NBr$ | | | | 171.1 | 4 | 15) |
| $Pr_4NI$ | 191 | 64 | 114) | 172.9 | 5 | 15) |
| | 190.7 | 162 | 109) | 172.3 | 7 | 115) |
| $Pr_4NPi$ | 156.2 | 27 | 67) | | | |
| $Bu_4NBPh_4$ | | | | 119.8 | 8 | 133) |
| $Bu_4NNO_3$ | 187.1 | 143 | 63) | | | |
| $Bu_4NPF_6$ | | | | 164.8 | 0 | 104) |
| $Bu_4NCl$ | 172 | 435 | 109) | | | |
| | 188 | 600 | 58) | | | |
| $Bu_4NClO_4$ | 182.8 | 80 | 63) | | | |
| $Bu_4NBr$ | 183.2 | 264 | 63) | 162.1 | 2 | 15) |
| | | | | 161.1 | 3 | 115) |
| $Bu_4NI$ | 180.2 | $6.1 \times 10^3$ | 69) | 163.7 | 3 | 115) |
| | 180.3 | 143 | 63) | 164.0 | 3 | 15) |
| $Bu_4NPi$ | 152.3 | 17 | 63) | 139.4 | 0 | 15) |
| $Bu_4Np$-toluene sulfonic acid | 152 | $4 \times 10^2$ | 58) | | | |
| $(n\text{-}Am)_4NBr$ | 174.4 | 220 | 109) | | | |
| $(n\text{-}Am)_4NPi$ | | | | 135.1 | 39 | 113) |
| $(i\text{-}Am)_3BuNBPh_4$ | | | | | | |
| $(i\text{-}Am)_3BuNI$ | | | | | | |
| $(i\text{-}Am)_3BuNPi$ | | | | 135.7 | 0 | 115) |
| $Hex_4NBr$ | | | | | | |
| $Hex_4NI$ | | | | | | |
| $Hept_4NI$ | | | | | | |

| Adiponitrile [142] | | | 1-Butanol | | | Dimethylacetamide [65] | | |
|---|---|---|---|---|---|---|---|---|
| $\Lambda_0$ | $K_A$ | Ref. | $\Lambda_0$ | $K_A$ | Ref. | $\Lambda_0$ | $K_A$ | Ref. |
| 13.06 | 0 | | 18.70 | $1.3 \times 10^3$ | [106] | | | |
| | | | 19.72 | $1.4 \times 10^3$ | [106] | 74.5 | | |
| 11.71 | 0 | | 17.01 | 920 | [106] | | | |
| 12.08 | 0 | | 18.12 | $1.2 \times 10^3$ | [106] | 67.9 | | |
| | | | 15.48 | 630 | [106] | | | |
| | | | 19.06 | $2.2 \times 10^3$ | [106] | | | |
| 10.94 | 0 | | 16.07 | 860 | [106] | | | |
| 11.30 | 0 | | 17.16 | $1.2 \times 10^3$ | [106] | 64.6 | | |
| 7.58 | 0 | | | | | | | |
| 10.91 | 0 | | 16.99 | $1.4 \times 10^3$ | [106] | | | |
| 10.06 | 0 | | | | | | | |
| 10.40 | 0 | | 15.57 | $1.3 \times 10^3$ | [106] | | | |

27

Table 2 (continued)

| Salt | Dimethylformamide | | | Dimethylsulfoxide | | |
|------|-------|-------|------|-------|-------|------|
| | $\Lambda_0$ | $K_A$ | Ref. | $\Lambda_0$ | $K_A$ | Ref. |
| $Me_4NCl$ | | | | | | |
| $Me_4NBr$ | 92.5 | 37 | 107) | 42.63 | | 31) |
| $Me_4NI$ | 90.9 | 14 | 107) | 42.40 | | 31) |
| $Me_4NPi$ | 76.4 | 0 | 57) | | | |
| $Me_3OctdNNO_3$ | | | | 37.0 | | 82) |
| $Me_3OctdNI$ | | | | 33.8 | | 82) |
| $Me_3PhNCl$ | 86.9 | 50 | 26) | | | |
| $Me_3PhNI$ | 84.2 | 10 | 26) | 37.8 | | 82) |
| $Et_4NClO_4$ | | | | 40.76 | | 81) |
| $Et_4NBr$ | 89.2 | 16 | 107) | 41.12 | | 31) |
| $Et_4NI$ | 87.5 | 12 | 107) | 40.86 | | 31) |
| $Et_4NPi$ | | | | | | |
| $Et_3NHBr$ | 89.1 | 333 | 57) | | | |
| $Et_3NHPi$ | 72.8 | 0 | 57) | | | |
| $EtNH_3Br$ | 91.8 | 125 | 57) | | | |
| $Pr_4NCl$ | | | | 37.83 | | 31) |
| $Pr_4NBr$ | 82.8 | 12 | 107) | 37.45 | | 31) |
| $Pr_4NI$ | 81.1 | 8 | 107) | 36.22 | | 31) |
| $Bu_4NBPh_4$ | | | | 22.02 | | 81) |
| $Bu_4NCl$ | | | | | | |
| $Bu_4NClO_4$ | | | | 35.31 | | 81) |
| $Bu_4NBr$ | | | | 35.65 | | 31) |
| $Bu_4NI$ | 77.7 | 8 | 107) | 35.39 | | 31) |
| | | | | 35.0 | | 82) |
| $(n\text{-}Am)_4NI$ | | | | 34.21 | | 31) |
| $(n\text{-}Am)_4NBPh_4$ | | | | 21.23 | | 31) |
| $(i\text{-}Am)_4NB(i\text{-}Am)_4$ | | | | 21.21 | | 31) |
| $(i\text{-}Am)_4NI$ | | | | 34.41 | | 31) |
| $(i\text{-}Am)_3BuNI$ | | | | | | |

| Ethanol | | | Ethylene carbonate [80] (40 °C) | | | Formamide [83] | | |
|---|---|---|---|---|---|---|---|---|
| $\Lambda_0$ | $K_A$ | Ref. | $\Lambda_0$ | $K_A$ | Ref. | $\Lambda_0$ | $K_A$ | Ref. |
| 51.87 | 141 | 116) | | | | | | |
| 51.67 | 122 | 117) | | | | | | |
| 54.03 | 164 | 116) | | | | | | |
| 53.56 | 146 | 117) | | | | | | |
| | | | 44.81 | 3 | | | | |
| 55.03 | 110 | 116) | | | | | | |
| | | | | | | 27.8 | | |
| | | | | | | 27.3 | | |
| 61.4 | 270 | 77) | 42.13 | | | | | |
| 53.54 | 96 | 120) | 42.48 | | | | | |
| 53.15 | 99 | 117) | | | | | | |
| 56.34 | 133 | 117) | 42.83 | | | | | |
| 56.5 | 130 | 120) | | | | | | |
| 54.2 | 69 | 121) | | | | | | |
| 51.2 | 100 | 77) | | | | | | |
| 46.86 | 78 | 117) | | | | | | |
| 49.94 | 120 | 117) | | | | | | |
| 41.54 | 39 | 117) | | | | | | |
| | | | 36.52 | | | | | |
| 43.51 | 75 | 117) | 36.97 | 4 | | | | |
| 46.65 | 123 | 117) | 37.41 | | | 3.4 | | |
| 45.31 | 130 | 117) | | | | | | |

Table 2 (continued)

| Salt | Dimethylformamide | | | Dimethylsulfoxide | | |
|------|-------------------|---|------|-------------------|---|------|
| | $\Lambda_0$ | $K_A$ | Ref. | $\Lambda_0$ | $K_A$ | Ref. |
| $Hex_4NI$ | | | | 33.61 | | 31) |
| $Hept_4NI$ | | | | 32.98 | | 31) |

| Salt | Methanol | | | N-Methylacetamide (40 °C) | | |
|------|----------|---|------|---------------------------|---|------|
| | $\Lambda_0$ | $K_A$ | Ref. | $\Lambda_0$ | $K_A$ | Ref. |
| $Me_4NCl$ | 120.82 | 7 | 118) | | | |
| $Me_4NBr$ | 125.16 | 14 | 118) | | | |
| $Me_4NI$ | 131.35 | 18 | 118) | 26.8 | | 127) |
| $Me_4NPi$ | 115.87 | 11 | 118) | | | |
| $Me_3OctdNNO_3$ | | | | 21.6 | | 66) |
| $Me_3OctdNI$ | | | | 21.8 | | 66) |
| $Et_4NBPh_4$ | | | | | | |
| $Et_4NNO_3$ | | | | | | |
| $Et_4NSCN$ | | | | | | |
| $Et_4NCl$ | | | | | | |
| $Et_4NClO_4$ | 131.39 | 41 | 119) | | | |
| $Et_4NBr$ | 116.95 | 10 | 118) | 21.3 (35 °C) | | 87) |
| $Et_4NI$ | | | | 26.2 | | 127) |
| $Et_4NPi$ | 107.31 | 18 | 122) | 21.3 (35 °C) | | 87) |
| | 107.63 | 13 | 123) | | | |
| $Et_4NAc$ | | | | | | |
| $(EtOH)_4NBr$ | 94.0 | 13 | 27) | | | |
| $(EtOH)_4NI$ | 99.95 | 12 | 27) | | | |
| $Pr_4NCl$ | | | | | | |
| $Pr_4NBr$ | 102.55 | 6 | 118) | | | |
| $Pr_4NI$ | 108.85 | 17 | 118) | 23.8 | | 127) |
| $Pr_4NPi$ | 93.12 | 21 | 122) | | | |
| $Bu_4NPBh_4$ | 76.00 | 37 | 84) | | | |
| $Bu_4NNO_3$ | | | | | | |
| $Bu_4NCl$ | 91.38 | 0 | 118) | | | |
| $Bu_4NBr$ | 95.39 | 3 | 118) | | | |

| Ethanol | | | Ethylene carbonate [80] (40 °C) | | | Formamide [83] | | |
|---|---|---|---|---|---|---|---|---|
| $\Lambda_0$ | $K_A$ | Ref. | $\Lambda_0$ | $K_A$ | Ref. | $\Lambda_0$ | $K_A$ | Ref. |
| 41.93 | 139 | [117] | | | | | | |

| Methylethylketone | | | Nitrobenzene | | | Nitromethane | | |
|---|---|---|---|---|---|---|---|---|
| $\Lambda_0$ | $K_A$ | Ref. | $\Lambda_0$ | $K_A$ | Ref. | $\Lambda_0$ | $K_A$ | Ref. |
| | | | | | | 116.44 | 45 | [125] |
| 161.8 | $4.5 \times 10^3$ | [126] | | | | 116.89 | 31 | [117] |
| 163.6 | 970 | [126] | | | | | | |
| | | | 33.3 | 24 | [128] | | | |
| | | | 34.0 | 20 | [90] | | | |
| | | | 34.51 | 40 | [90] | 114.3 | | [89] |
| | | | | | | 119.7 | | [89] |
| | | | 38.5 | 80 | [90] | 110.06 | 2 | [125] |
| | | | 37.6 | | [92] | 113.4 | | [89] |
| 158.3 | 960 | [126] | 33.48 | 62 | [90] | 110.45 | 2 | [125] |
| 159.7 | 411 | [130] | | | | 111.2 | | [89] |
| | | | 32.4 | 7 | [128] | 93.5 | | [89] |
| | | | 32.7 | | [92] | | | |
| | | | 35.5 | 150 | [90] | | | |
| | | | | | | 101.61 | | [125] |
| 146.4 | 940 | [126] | | | | 102.13 | | [125] |
| 148.7 | 440 | [126] | | | | | | |
| | | | 29.5 | 3 | [128] | | | |
| 102.8 | 0 | [126] | 22.34 | 5 | [52] | | | |
| | | | 34.34 | 24 | [114] | | | |
| | | | | | | 96.58 | | [125] |
| 139.5 | 790 | [126] | 33.5 | 56 | [90] | 96.97 | | [125] |

Table 2 (continued)

| Salt | Methanol | | | $N$-Methylacetamide (40 °C) | | |
|---|---|---|---|---|---|---|
| | $\Lambda_0$ | $K_A$ | Ref. | $\Lambda_0$ | $K_A$ | Ref. |
| Bu$_4$NI | 101.72 | 16 | 118) | 22.5 | | 127) |
| Bu$_4$NPi | 86.10 | 16 | 84) | | | |
| | 86.14 | 7 | 118) | | | |
| | 86.04 | 13 | 122) | | | |
| ($n$-Am)$_4$NBr | 91.41 | 2 | 118) | | | |
| ($n$-Am)$_4$NI | 97.42 | 16 | 118) | 22.0 | | 127) |
| ($n$-Am)$_4$NSCN | | | | | | |
| ($i$-Am)$_4$NI | 98.04 | 13 | 124) | | | |
| ($i$-AM)$_4$NPi | 82.54 | 15 | 122) | | | |
| ($i$-Am)$_3$BuNBPh$_4$ | 73.3 | 23 | 84) | | | |
| ($i$-Am)$_3$BuNI | 99.39 | 17 | 84) | | | |
| ($i$-Am)$_3$BuNPi | 83.69 | 10 | 84) | | | |
| Hex$_4$NI | | | | 21.8 | | 127) |
| Hept$_4$NI | | | | 21.5 | | 127) |

| Salt | 1-Propanol | | | Propylene carbonate | | |
|---|---|---|---|---|---|---|
| | $\Lambda_0$ | $K_A$ | Ref. | $\Lambda_0$ | $K_A$ | Ref. |
| Me$_4$NCl | 25.05 | 456 | 117) | | | |
| Me$_4$NClO$_4$ | | | | | | |
| Me$_4$NBr | 26.91 | 638 | 117) | | | |
| Me$_4$NPi | | | | | | |
| Me$_3$EtNPi | | | | | | |
| Et$_4$NClO$_4$ | | | | 32.06 | | 101) |
| Et$_4$NBr | 27.19 | 373 | 117) | | | |
| | 27.07 | 393 | 129) | | | |
| Et$_4$NI | 29.01 | 466 | 117) | | | |
| | 28.52 | 503 | 129) | | | |
| Et$_4$NPi | | | | | | |
| Pr$_4$NClO$_4$ | | | | | | |
| Pr$_4$NBr | 24.42 | 270 | 117) | | | |
| | 24.43 | 311 | 129) | | | |

| Methylethylketone | | | Nitrobenzene | | | Nitromethane | | |
|---|---|---|---|---|---|---|---|---|
| $\Lambda_0$ | $K_A$ | Ref. | $\Lambda_0$ | $K_A$ | Ref. | $\Lambda_0$ | $K_A$ | Ref. |
| 141.7 | 380 | 126) | 32.80 | 27 | 114) | | | |
| | | | 27.83 | 7 | 114) | | | |
| | | | 27.9 | 3 | 128) | | | |
| 134.8 | 760 | 126) | | | | | | |
| 136.9 | 350 | 126) | | | | | | |
| | | | 33.26 | 37 | 131) | | | |
| 133.3 | 330 | 126) | | | | | | |
| 130.5 | 310 | 126) | | | | | | |

| Pyridine | | | Sulfolane (30 °C) | | | Misc. | | |
|---|---|---|---|---|---|---|---|---|
| $\Lambda_0$ | $K_A$ | Ref. | $\Lambda_0$ | $K_A$ | Ref. | $\Lambda_0$ | $K_A$ | Ref. |
| | | | 10.99 | | 14) | | | |
| 76.7 | $1.5 \times 10^3$ | 96) | | | | | | |
| 75.5 | $1.2 \times 10^3$ | 96) | | | | | | |
| 90.8 | 156 | 119) | 10.63 | | 14) | valeronitrile 88.3 | 194 | 132) |
| | | | | | | N-methylformamide 47.7 | | 88) |
| | | | 11.20 | 5 | 95) | | | |
| | | | | | | N-methylformamide 39.3 | | 88) |
| | | | 9.91 | | 14) | | | |

33

Table 2 (continued)

| Salt | 1-Propanol | | | Propylene carbonate | | |
|------|-----------|---|------|--------------------|---|------|
| | $\Lambda_0$ | $K_A$ | Ref. | $\Lambda_0$ | $K_A$ | Ref. |
| $Pr_4NI$ | 26.08 | 391 | 117) | | | |
| | 25.80 | 418 | 129) | | | |
| $Bu_4NBPh_4$ | | | | 17.14 | | 52) |
| $Bu_4NNO_3$ | | | | | | |
| $Bu_4NCl$ | 21.16 | 149 | 117) | | | |
| $Bu_4NClO_4$ | 27.13 | 769 | 117) | 28.17 | | 101) |
| $Bu_4NBr$ | 22.92 | 266 | 117) | 28.65 | | 94) |
| $Bu_4NI$ | 24.60 | 415 | 117) | | | |
| $Bu_4NPi$ | | | | | | |
| $(i\text{-}Am)_4NB(i\text{-}AM)_4$ | | | | 16.37 | | 101) |
| $(i\text{-}Am)_4NI$ | | | | 26.95 | | 101) |
| $(i\text{-}Am)_3BuNBPh_4$ | | | | | | |
| $(i\text{-}Am)_3BuNI$ | 24.02 | 462 | 117) | | | |
| $Hept_4NI$ | 22.18 | 442 | 117) | | | |
| $EtPh_3AsPi$ | | | | | | |

| Pyridine | | | Sulfolane (30 $^\circ$C) | | | Misc. | | |
|---|---|---|---|---|---|---|---|---|
| $\Lambda_0$ | $K_A$ | Ref. | $\Lambda_0$ | $K_A$ | Ref. | $\Lambda_0$ | $K_A$ | Ref. |
| | | | | | | tetrahydrofuran | | |
| | | | | | | 84.8 | $2.3 \times 10^4$ | [97] |
| | | | | | | isobutyronitrile | | |
| | | | | | | 81.61 | 18 | [133] |
| 76.6 | $2.7 \times 10^3$ | [96] | | | | | | |
| | | | 9.49 | | [14] | | | |
| 75.3 | $4 \times 10^3$ | [102] | | | | 1-pentanol | | |
| | | | | | | 11.31 | $2.5 \times 10^3$ | [106] |
| 73.1 | $2.4 \times 10^3$ | [102] | | | | 1-pentanol | | |
| | | | | | | 12.00 | $3.2 \times 10^3$ | [106] |
| 57.7 | 780 | [102] | | | | | | |
| | | | | | | tetrahydrofuran | | |
| | | | | | | 80.6 | $1.7 \times 10^4$ | [97] |
| | | | | | | 1-pentanol | | |
| | | | | | | 11.62 | $3.3 \times 10^3$ | [106] |
| | | | | | | 1-pentanol | | |
| | | | | | | 10.76 | $3.2 \times 10^3$ | [106] |
| 57.7 | 510 | [102] | | | | | | |

Table 3. *Limiting ionic conductivities of cations in selected organic solvents*
All values measured at 25 °C unless otherwise indicated

| Ion | $\lambda_i^0$ | | | |
|---|---|---|---|---|
| | Acetone[15] | Acetonitrile | Dimethyl-acetamide | Dimethyl-formamide |
| $H^+$ | | | | |
| $Li^+$ | 72.8 [63) | 69.3 [50) | | 25.0 [9) |
| $Na^+$ | 78.4 [63) | 76.9 [50) | 25.6 [9) | 29.9 [9) |
| $K^+$ | 80.6 [63) | 83.6 [50) | 25.2 [9) | 30.8 [9) |
| $Rb^+$ | | 85.6 [50) | | 32.4 [9) |
| $Cs^+$ | | 87.3 [50) | | 34.5 [9) |
| $Cu^+$ | | 64.7 [10) | | |
| $Ag^+$ | | 86.0 [10) | | |
| $Tl^+$ | | 91.5 [70) | | |
| $NH_4^+$ | 94.5 [109) | | | 38.7 [9) |
| $Me_4N^+$ | 97.7 [63) | 94.5 [50) | | 38.9 [107) |
| $Me_3OctdN^+$ | | | | |
| $Me_3PhN^+$ | | | | 31.9 [26) |
| $Et_4N^+$ | 89.0 [63) | 84.8 [50) | 32.7 [65) | 35.6 [107) |
| $(EtOH)_4N^+$ | | 64.0 [50) | | |
| $(n\text{-}Pr)_4N^+$ | 77.7 [63) | 70.3 [50) | 26.2 [65) | 29.2 [107) |
| $(n\text{-}Bu)_4N^+$ | 67.3 [63) | 64.1 [50) | 22.8 [65) | 25.4 [107) |
| | | | | 26.2 [9) |
| $(n\text{-}Am)_4N^+$ | 58.8 [109,15) | 56.0 [124) | | |
| $(i\text{-}Am)_4N^+$ | | 56.8 [124) | | |
| $(i\text{-}Am)_3BuN^+$ | | 58.1 [124) | | |
| $Hex_4N^+$ | | | | |
| $Hept_4N^+$ | | | | |
| $Ph_4As^+$ | | 55.8 [50) | | |

| | | $\lambda_i^0$ | |
|---|---|---|---|
| Dimethyl-sulfoxide | Ethanol | Ethylene carbonate [80] (40 °C) | Formamide [138] |
| 14.5 [76] | | | |
| 17.1 [136] | | | 10.8 |
| 11.4 [137] | 17.05 [117] | 7 | 8.5 |
| 13.5 [81] | 20.31 [117] | 13 | 10.1 |
| 13.8 [82] | | | |
| 13.1 [9] | | | |
| 14.7 [81] | | | |
| 14.4 [82] | | | |
| 13.9 [9] | 23.55 [117] | 15 | 12.7 |
| | | 16 | 12.8 |
| | | 17 | 13.5 |
| | 29.65 [117] | 18 | 12.5 |
| 10.0 [82] | | | |
| 14.1 [82] | | | 10.7 |
| 17.06 [31] | 29.27 [117] | 15.8 | 10.0 |
| 16.5 [81] | | | |
| 13.42 [31] | 22.98 [117] | | |
| 11.59 [31] | 19.67 [117] | 10 | 6.8 |
| 11.0 [81] | | | |
| 11:2 [9,82] | | | |
| 10.41 [31] | | | |
| 10.61 [31] | | | |
| | 18.31 [117] | | |
| 9.79 [31] | | | |
| 9.18 [31] | 14.93 [117] | | |

Table 3 (continued)

| Ion | Methanol | N-Methyl-acetamide (40 °C) | Nitro-benzene | Nitro-methane |
|---|---|---|---|---|
| | | $\lambda_i^0$ | | |
| $H^+$ | | 9.1 [66] | | 63 [89] |
| $Li^+$ | 39.6 [9] | 6.6 [66] | | 55 [89] |
| $Na^+$ | 45.21 [84]<br>45.2 [9] | 8.2 [66] | 16.3 [90] | 58 [89] |
| $K^+$ | 52.38 [84]<br>52.4 [9] | 8.4 [66] | 17.8 [90] | 60 [89] |
| $Rb^+$ | | | | |
| $Cs^+$ | 60.83 [139] | | | |
| $Ag^+$ | | | | 52 [89] |
| $Tl^+$ | | | | 60 [89] |
| $NH_4^+$ | | 9.7 [66] | 18.4 [90] | 64 [89] |
| $Me_4N^+$ | 68.7 [118]<br>66.7 [122] | 12.0 [66]<br>12.1 [127] | 17.1 [140] | 54.51 [125] |
| $Me_3EtN^+$ | | | | |
| $Me_3OctdN^+$ | | 7.1 [66] | | |
| $Me_3PhN^+$ | | 10.2 [66] | | |
| $Et_4N^+$ | 60.5 [118]<br>58.2 [122] | 11.6 [66]<br>11.6 [127] | 16.4 [140] | 47.60 [125] |
| $(EtOH)_4N^+$ | 37.3 [27] | | | |
| $(n\text{-}Pr)_4N^+$ | 46.08 [118]<br>43.9 [122] | 9.1 [66]<br>9.1 [127] | 13.3 [140] | 39.14 [125] |
| $(i\text{-}Pr)_4N^+$ | | | | 34.06 [125] |
| $(n\text{-}Bu)_4N^+$ | 38.94 [118]<br>36.9 [122] | 7.8 [66]<br>7.8 [127] | 11.9 [140] | |
| $(n\text{-}Am)_4N^+$ | 34.8 [124] | 7.3 [127] | | |
| $(i\text{-}Am)_4N^+$ | 35.3 [124] | | | |
| $(i\text{-}Am)_3BuN^+$ | 36.4 [124] | | | |
| $Hex_4N^+$ | | 7.1 [127] | | |
| $Hept_4N^+$ | | 6.8 [127] | | |

| 1-Propanol[117] | $\lambda_i^0$ Propylene-carbonate [101] | Pyridine [91] | Sulfolane (40 °C) | |
|---|---|---|---|---|
| | 7.3 | 24.9 | 4.33 | [141] |
| | | 26.8 | 3.61 | [141] |
| | 11.97 | 32.0 | 4.05 | [141] |
| | | | 4.16 | [141] |
| | | | 4.27 | [141] |
| | | 34.3 | | |
| | | 46.8 | 4.97 | [141] |
| 14.40 | | 43.0 | 4.31 | [14] |
| | | 41.8 | | |
| 15.05 | 13.28 | | 3.95 | [14] |
| 12.19 | | | 3.23 | [14] |
| 10.71 | 9.39 | 24.0 | 2.80 | [14] |
| 10.17 | 8.18 | | | |
| 8.29 | | | | |

Table 3 (continued)

| Ion | Acetone[15,63] | Acetonitrile | Dimethyl-acetamide | Dimethyl-formamide[9] |
|---|---|---|---|---|
| $F^-$ | 85.0 | | | |
| $Cl^-$ | 105.2 | 98.4 [105] | | 55.1 |
| $Br^-$ | 115.9 | 100.7 [50] | 43.2 [9] | 53.6 |
| $I^-$ | 113.0 | 102.4 [50] | 41.8 [9] | 52.3 |
| $SCN^-$ | | 113.4 [70] | 48.8 [65] | 59.8 |
| $NO_3^-$ | 120.1 | 106.4 [10] | 46.3 [65] | 57.3 |
| $BF_4^-$ | | 108.5 [10] | | |
| $BBr_4^-$ | 113.0 | | | |
| $PF_6^-$ | | 104.2 [10] | | |
| $ClO_4^-$ | 115.3 | 103.7 [50] | 42.8 [65] | 52.4 |
| | | | 43.0 [9] | |
| $Ph_4B^-$ | | 58.3 [50] | | |
| $Ph_3ClB^-$ | | 114.5 [110] | | |
| $(i\text{-}Am)_4B^-$ | | 57.6 [50] | | |
| $HCOO^-$ | | | | |
| $CH_3COO^-$ | | | | |
| $PhCOO^-$ | | | | |
| $PhSO_3^-$ | | | 31.0 | |
| $OctdSO_4^-$ | | | | |
| $Pi^-$ | 85.3 | 77.7 [50] | 31.5 | 38.1 |
| | | | | 31.5 [107] |

| Dimethyl-sulfoxide | | Ethanol [117] | $\lambda_i^0$ Ethylene carbonate [80] (40 °C) | Formamide [138] |
|---|---|---|---|---|
| 23.9 | [137] | 21.87 | | 17.1 |
| 24.40 | [31] | | | |
| 24.06 | [31] | 23.88 | 26.5 | 17.2 |
| 24.2 | [82] | | | |
| 24.7 | [9] | | | |
| 23.80 | [31] | 27.00 | 27 | 16.6 |
| 23.8 | [82] | | | |
| 24.3 | [9] | | | |
| 29.2 | [82] | | | 17.2 |
| 27.0 | [82] | | | 17.4 |
| 24.6 | [82] | | 26 | |
| 24.3 | [81] | | | |
| 25.2 | [9] | | | |
| 10.61 | [31] | | | |
| 11.0 | [81] | | | |
| 10.61 | [31] | | | |
| | | | | 15.3 |
| | | | | 11.9 |
| | | | | 9.8 |
| 16.8 | [82] | | | 10.4 |
| 10.0 | [82] | | | |
| 17.3 | [82] | | | |

Table 3 (continued)

| Ion | Methanol Methanol | | $\lambda_i^0$ N-Methyl-acetamide[66] (40 °C) | Nitro-benzene [90] | Nitro-methane | |
|---|---|---|---|---|---|---|
| Cl⁻ | 52.36 | [118] | | | | |
| | 54.7 | [76] | 11.5 | 22.2 | 62.52 | [125] |
| Br⁻ | 56.45 | [118] | | | | |
| | 56.5 | [9] | 12.8 | 21.6 | 62.88 | [125] |
| I⁻ | 62.78 | [118] | 14.6 | 20.4 | 62 | [89] |
| | 62.8 | [9] | | | | |
| SCN⁻ | 64.7 | | 16.1 | | 70 | [89] |
| NO₃⁻ | | | 14.5 | 22.6 | 64 | [89] |
| PF₆⁻ | | | | | | |
| ClO₄⁻ | 71.0 | [9] | 16.8 | 20.9 | 64 | [89] |
| Ph₄B⁻ | 36.50 | [134] | | | | |
| Ph₃FB⁻ | | | | | | |
| (i-Am₄)B⁻ | | | | | | |
| BrO₃⁻ | | | 13.6 | | | |
| CH₃COO⁻ | | | | | | |
| PhSO₃⁻ | | | 10.3 | | | |
| OctdSO₄⁻ | | | 7.1 | | | |
| Pi⁻ | 47.14 | [118] | 11.8 | 16.0 | 44 | [89] |
| | 49.2 | [122] | | | | |

## B. Correlations

### 1. General

Ion pair association constants measured by conductance have not often been verified by independent techniques, but where comparisons have been made agreement appears satisfactory. For example, a value of 70.2 ± 0.5 was obtained for the association constant of silver nitrate in acetonitrile by conductance [10], and a value of 74 ± 5 by potentiometric measurements [144] of the cell

$$Ag \,|x \text{ M AgNO}_3 \,\|\, 0.1 \text{ M Et}_4\text{NClO}_4 \,\|\, x \text{ M AgClO}_4 \,|Ag$$

Interpretation of ion mobility trends begins with Stokes Law

| 1-Propanol[117] | Propylene carbonate[101] | $\lambda_i^0$ Pyridine | | Sulfolane [141] (40 °C) |
|---|---|---|---|---|
| 10.45 | 20.20 | | | 9.30 |
| 12.22 | 19.26 | 51.3 | [102] | 8.92 |
| 13.81 | 18.76 | 49.1 | [102] | 7.22 |
| | | 48.4 | [91] | |
| | | | | 9.64 |
| | | 52.6 | [91] | |
| | | | | 5.95 |
| 15.42 | 18.78 | 47.6 | [102] | 6.68 |
| | | 24.0 | [91] | |
| | 8.18 | | | |
| | | 52 | [91] | |
| | | 33.7 | [102] | |

$$\lambda_0 \eta_0 = \frac{|Z|F^2}{6\pi Nr}$$

where $r$ is the radius of the ion in the solvent medium. In the absence of the factors affecting ion-solvent interactions mentioned above, Stokes Law predicts a linear relationship between ion mobility and the reciprocal of this radius. This relationship is a convenient point of departure for discussions of ionic mobilities. A modification of this expression to account for a retarding effect on mobility due to solvent relaxation has been proposed by Zwanzig [149]. Use of this theory enables reasonable predictions of mobility for monovalent ions in some solvents [150].

Data of sufficient precision to be treated by the Fuoss-Onsager conductance expression yield, in addition to values for $\Lambda_0$ and $K_A$, an ion size parameter term $\overset{\circ}{a}$ which is defined as the internuclear distance of closest approach. This term gives reasonable values on the order of 3 to 6 Å; however discernible trends in a series of similar salts are not observed. For example, $\overset{\circ}{a}$ values for the tetraalkylammonium bromides, iodides, and picrates in acetonitrile are all about the same, 3.6 ± 0.2, except for tetramethylammonium bromide, which is 4.4 [15]. A similar situation is found for the tetramethyl- through tetrabutyl ammonium chlorides in nitromethane, where $\overset{\circ}{a}$ is 3.9 ± 0.3 if the same viscosity correction used in acetonitrile is employed [125].

In addition, the size of $\overset{\circ}{a}$ is very dependent on the precision of the data, as shown by comparison of $\overset{\circ}{a}$ values for slightly associated salts. An increase in $\overset{\circ}{a}$ of 10% may be seen in acetonitrile when $K_A$ is changed from 0 to 1 or 2, and a $K_A$ of about 5 may change $\overset{\circ}{a}$ by an angstrom over the value found when no association is assumed. Evans, Zawoyski and Kay point out that in the case of $Me_4NBr$, a reduction in the $K_A$ value from 46 to 43 gives a decrease in $\overset{\circ}{a}$ from 4.4 to 3.6 units[15]. Thus, for associated salts at least, the relationship between ion size(or ion-ion contact distance) and $\overset{\circ}{a}$ cannot be interpreted with any confidence. Springer, Coetzee and Kay [50] have discussed the use of accurately measured transference numbers in conjunction with conductance measurements to obtain an estimate of the electrophoretic contribution to the mobility of an ion. They found that for tetramethylammonium perchlorate in acetonitrile an ion size parameter of 8 Å gave good fit of the data to theory for the Onsager limiting law after expansion to include an ion size term and elimination of concentration terms of power 3/2 or greater. This value is more reasonable than the value of 3.1 obtained by conductance. They conclude that the low $\overset{\circ}{a}$ values should be attributed to approximations in evaluation of the relaxation effect made in the Fuoss-Onsager conductance equation.

## 2. Conductance Results in Acetonitrile

Conductance in acetonitrile has been extensively studied, so for this reason, and because it has a moderate dielectric constant that fosters some ion-pairing, it provides a good example of the conclusions that can be drawn from conductance data.

Limiting single ion conductivities in acetonitrile have been plotted against reciprocal crystallographic radii in Fig. 3. It is seen that for large ions conductance is, as expected from Stokes law, largely a function of size. However, with smaller ions, solvation becomes important and a reverse size-mobility dependence appears. In addition, cations and anions begin showing different size dependencies on mobility. This can be interpreted in terms of the ability of acetonitrile to solvate cations to a greater extent than anions, for actonitrile, though a weak Lewis base, is an even weaker Lewis acid. Therefore anions become relatively more mobile than cations.

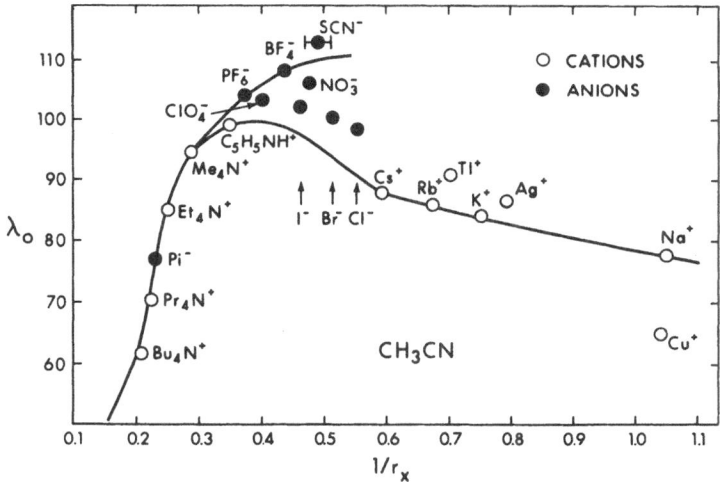

Fig. 3. Single ion conductivities in acetonitrile *vs.* reciprocal of estimated crystallographic radii

Table 4. *Association constants for some 1 : 1 electrolytes in acetonitrile*[a]

|  | $Cl^-$ | $Br^-$ | $SCN^-$ | $NO_3^-$ | $I^-$ | $BF_4^-$ | $ClO_4^-$ | $PF_6^-$ | $Pi^-$ | $BPh_4^-$ |
|---|---|---|---|---|---|---|---|---|---|---|
| $Me_4N^+$ | 56 | 46 |  |  | 19 |  | 7 | 5 | 1 | 0 |
| $Et_4N^+$ |  |  | 5 | 5 |  | 0 |  |  | 10 | 0 |
| $n\text{-}Pr_4N^+$ |  | 4 |  | 5 |  |  |  |  |  | 0 |
| $n\text{-}Bu_4N^+$ |  | 2 |  | 3 |  |  | 0 |  | 0 | 0 |
| $n\text{-}Am_4N^+$ |  |  |  |  |  |  |  |  | 39 | 0 |
| $Li^+$ |  |  |  |  |  | 4 |  |  |  |  |
| $Na^+$ |  |  | 87 | 0 |  |  | 10 |  |  | 0 |
| $K^+$ |  |  | 26 | 0 |  |  | 14 |  |  | 0 |
| $Rb^+$ |  |  |  |  |  |  | 19 |  |  | 0 |
| $Cs^+$ |  |  |  |  |  |  | 23 |  |  | 2 |
| $Cu^+$ |  |  | 0 |  | 9 |  | 0 | 15 |  |  |
| $Ag^+$ |  |  | 71 |  | 0 |  | 0 | 0 |  |  |
| $Tl^+$ |  |  |  |  | 14 |  | 32 |  |  |  |

[a] Anions arranged in order of increasing crystallographic radius.

Association trends also reflect for the most part increasing solvation with decreasing size (Table 4). Tetraalkylammonium salts become less associated as the size of the alkyl group becomes larger, an expected trend for these relative-

45

ly unsolvated cations. The opposite size dependence on association is found for the alkali metal perchlorates. These salts have a range of association constants from 4 to 20, with the most solvated ion studied, lithium, having the least amount of association. Lithium ion therefore appears to have the largest effective size of the alkali metal ions in acetonitrile. This explanation cannot be used to correlate association of the tetramethylammonium halide salts with relative halide mobilities, however.

Copper (I), silver (I), and thallium (I) mobilities deviate from the curve drawn through the alkali metal ions. Copper (I) is extensively solvated by acetonitrile, and the low mobility of this ion relative to the alkali metal ions can be related to this solvation. The high mobility of silver (I), which also undergoes strong interaction, is more difficult to explain. Silver (I) coordinates strongly to two acetonitrile molecules to give a linear species (this coordination is retained even when the silver is ion-paired to nitrate [145]). It may be that strong specific solvation to two acetonitrile molecules results in overall solvation that is less than that of the alkali metal trend, and results in higher mobility. In contrast, thallium (I) does not exhibit specific solvation by acetonitrile and yet it also has a high mobility. Therefore these mobilities cannot be related in a simple way to the extent of ion solvation.

Association constants for salts of copper, silver, and thallium appear to reflect solvation in a fairly simple way. For example, of the perchlorate salts, only those of the poorly solvated thallium ion show association.

Perchlorate and nitrate, as well as the halides, are less mobile relative to their size than the tetrafluoroborate and hexafluorophosphate ions. Anions that lie below the upper curve in Fig. 3 appear to undergo interaction with acetonitrile to a greater extent than would be expected on the basis of simple charge-dipole interaction, but the nature of this interaction is difficult to postulate. The fluoro ions should less likely undergo specific interactions. Because the differences in mobilities are not great, however, extensive speculation is unwarranted.


## 3. Comparisons of Conductance Data among Solvents

As mentioned, ion mobilities in a solvent afford some measure of the relative degree of ion solvation. Processes such as solvent structure, hydrogen-bonding, and Lewis acid-base interactions all affect net mobilities, however, and as a result comparisons among solvents are not as straightforward as comparisons within a single solvent. In the following section some general correlations and trends among mobilities and association will be dicussed.

Plots of limiting ion conductivities versus the reciprocal of the Pauling crystallographic radii for the alkali metal ions in various solvents are shown in Fig. 4 and 5. The important gross features are that mobilities increase with increas-

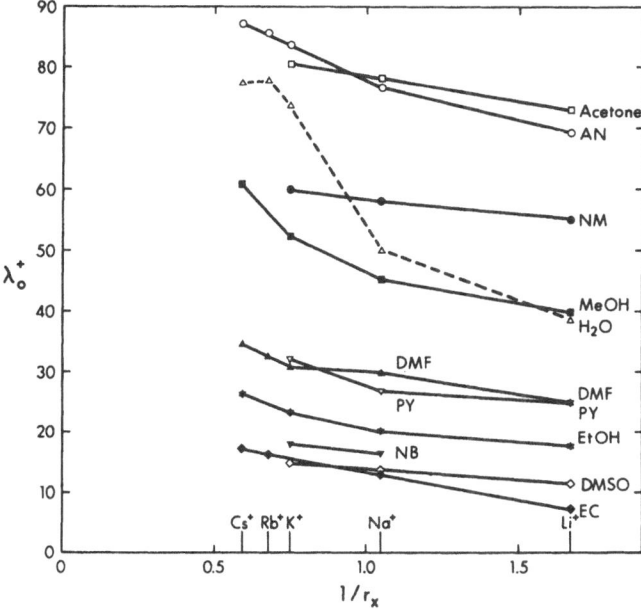

Fig. 4. Limiting single ion mobilities *vs.* reciprocal of estimated crystallographic radii of alkali metal ions: AN, acetonitrile; NM, nitromethane; DMF, dimethylformamide; PY, pyridine; NB, nitrobenzene; DMSO, dimethylsulfoxide; EC, ethylene carbonate

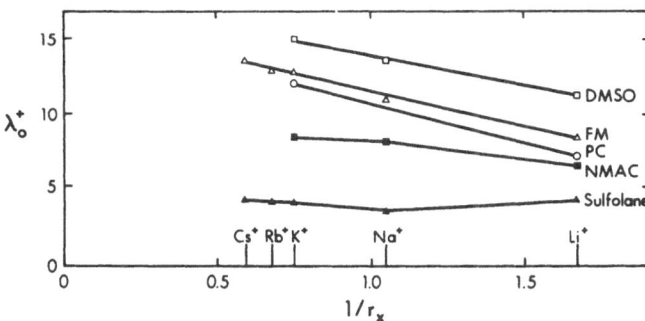

Fig. 5. Limiting single ion mobilities *vs.* reciprocal of estimated crystallographic radii of alkali metal ions in several viscous solvents: FM, formamide; PC, propylene carbonate; NMAC, *N*-methylacetamide

ing crystallographic radius, and that mobilities tend to be much higher in solvents of low viscosity. The mobility increase with ion size indicates that for these solvents, most of which have appreciable Lewis base character, solvation increases with decreasing ion size. Thus, the intense charge field around the

47

small lithium ion results in a highly solvated and therefore less mobile species. The only exception to this trend is the apparent high mobility for lithium ion in sulfolane, which may be the result of experimental difficulties.

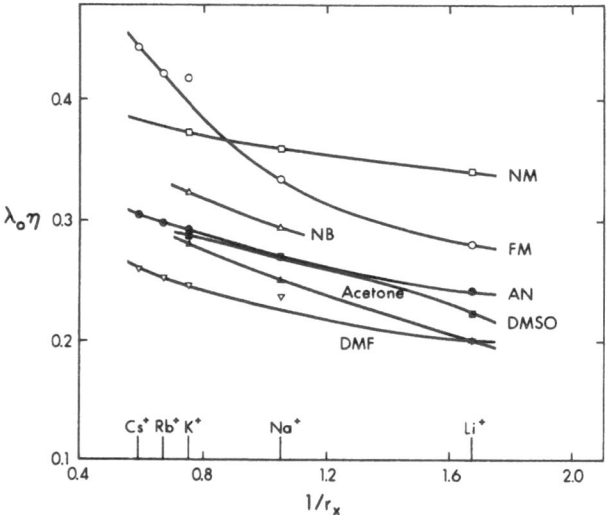

Fig. 6. Limiting single ion mobility-viscosity product *vs.* reciprocal of estimated crystallographic radii of alkali metal ions

To minimize the effects of viscosity for purposes of comparing data between solvents, plots are often made using the product of the ion mobility and the viscosity (Walden product) in place of mobility alone. A plot of the Walden product against the reciprocal of the crystallographic radii for several solvents is shown in Fig. 6. Arbitrary curves have been drawn to indicate general trends. Values in solvents for which precise transference numbers and conductance data are available, such as acetonitrile and nitromethane, give smooth curves.

The tetraalkylammonium ions form a series of large, relatively ideal cations that show fairly small solvation effects (with the possible exception of the tetramethylammonium ion) in most solvents. The Walden product increases with decreasing size as predicted by Stokes law (Fig. 7). These ions can therefore be used in assessing electrostatic interactions with minimum complications from specific nonelectrostatic ion-ion or ion-solvent interactions. The tetramethylammonium ion is the most mobile, though its mobility falls below the trend of the others in the series. Apparently its charge density is sufficiently high to cause ion-solvent dipole interactions to become important in most solvents. The tetramethylammonium ion is in many solvents more mobile than ions of higher charge density because the intense charge field around ions much smaller than it cause an increase in solvation and correspondingly lower mobil-

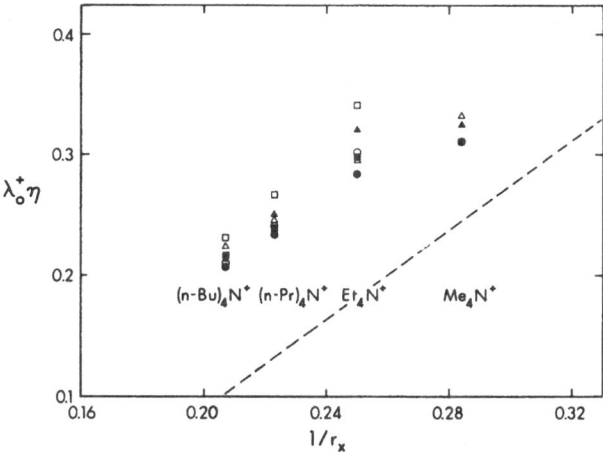

Fig. 7. Limiting single ion mobility-viscosity product *vs.* reciprocal of estimated crystallographic radii of tetraalkylammonium ions in various solvents. Stokes law plot shown as dashed line. □, dimethylsulfoxide; ■, nitrobenzene; △, acetonitrile; ▲, ethanol; ○, dimethylacetamide; ●, dimethylformamide

ities. High mobility is not a critical function of ion size, however, for the ammonium ion, of much smaller crystallographic radius, has a mobility close to that of the tetramethylammonium ion in a number of solvents. The mobility of the ammonium ion is much higher than that of the potassium ion in all solvents where data are available, although it has about the same crystallographic radius. Apparently the potassium ion is capable of being solvated to a greater extent than the ammonium ion.

A comparison of the limiting ionic conductance-viscosity products for cations in methanol, ethanol, and acetonitrile shows some interesting relationships (Fig. 8). The tetraalkylammonium ions are relatively more mobile in methanol and in ethanol than in acetonitrile. This may be a result of the appreciably greater solvent structuring present in the alcohols, or of the large dipole moment of acetonitrile compared to methanol and ethanol (4.0 *vs.* 1.7 debyes). The decreased mobilities of the alkali metal ions in ethanol relative to methanol reflects the larger size of the ethanol molecule. A corresponding effect on anions was pointed out earlier for the halide ions. The mobility of lithium and sodium ions in acetonitrile is greater than in methanol, but that for potassium is about the same in the two solvents, and that for cesium is lower in acetonitrile. It was suggested by Kay, Hales and Cunningham [16] that the crossover is the result of two competing solvation mechanisms, one due to dipole moment effects, the other caused by the greater Lewis base properties of the alcohols. For small cations solvation is primarily determined by the Lewis basicity of the solvent, but as the size of the cation increases and the distance of closest approach of the cation and solvent molecule becomes greater, charge-dipole interactions predominate.

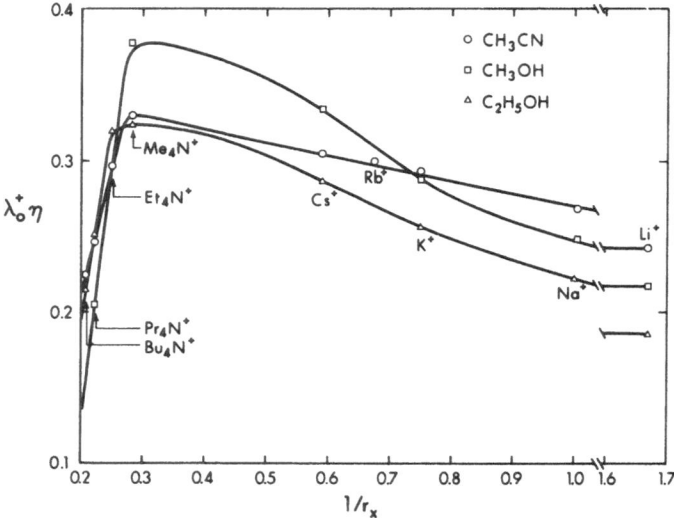

Fig. 8. Limiting single ion mobility-viscosity product *vs.* reciprocal of estimated crystallographic radii for cations in methanol, ethanol, and acetonitrile (from Ref. [16])

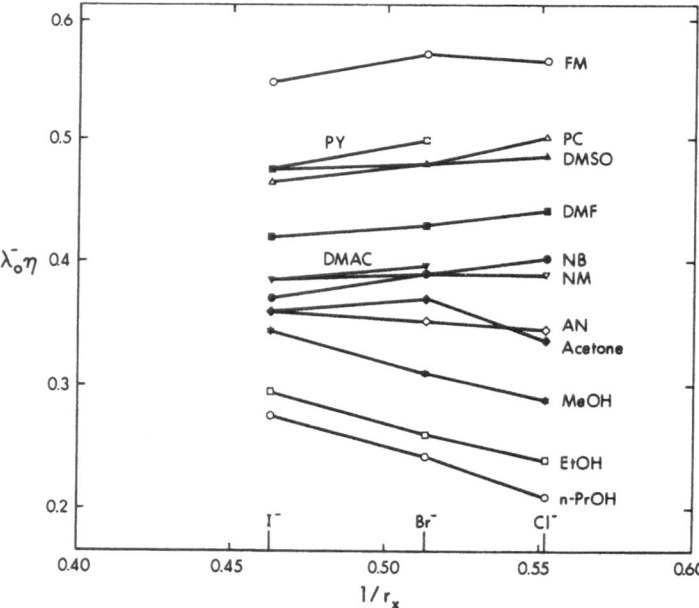

Fig. 9. Limiting single ion mobility-viscosity product *vs.* reciprocal of crytallographic radii for halide ions in various solvents

Halide ion mobilities follow the expected trends (Fig. 9). Chloride ion, with the smallest crystallographic radius of the three halides considered, is the most mobile in solvents such as nitrobenzene and dimethylformamide, where anion solvation is expected to be small. In these solvents the Walden products are large.

Hydroxylic solvents are capable of solvating anions through hydrogen bonding, and so halide mobilities are relatively low in alcohols, with chloride the least mobile. The mobility decreases observed for all the halides upon going up the homologous series of aliphatic alcohols may be the result of the increased size and mass of the alkyl group. A similar mass effect may be seen in the lowered mobility of the halides in dimethylacetamide compared to dimethylformamide. Here, as in the alcohol series, dipole moments and viscosities of the two solvents do not appear to be sufficiently different to explain the mobility differences.

The relatively high mobilities of bromide and iodide in pyridine may be the result of poor anion solvation by the diffuse positive end of the pyridine dipole. The other aryl solvent shown in Fig. 9, nitrobenzene, appears to solvate somewhat more readily, though the nature of anion interaction is not clear.

In formamide, acetone, and nitromethane the bromide ion is the most mobile of the halides. The difference is slight in nitromethane, but pronounced in the other two solvents. Because mobilities reflect a variety of factors it is possible that opposing effects could result in an ion of intermediate size being more mobile than others in the series. Another possible factor could be the presence of impurities in formamide and acetone, formamide because of decomposition on standing even a short time, and acetone because of the difficulty in removing last traces of water. The presence of impurities could have a significant but unpredictable effect on mobilities.

Acetone is the only solvent in which the conductance of a fluoride salt has been reported ($\lambda_0^- = 85.0$, $\lambda_0^- \eta = 0.27$, $1/r_{F^-} = 0.74$). The low mobility is as expected for a small, highly solvated ion. Data on fluoride mobilities in other solvents would be very useful.

As has been previously mentioned, ion-solvent interactions involve several factors that include electrostatic and Lewis acid-base contributions. It would be helpful to be able to estimate the magnitude of the individual contributions from these separate factors, and much effort is being directed toward devising ways of estimating the polarity of solvents. Several methods have been proposed. One empirical approach to the problem involves use of the relation

$$-\Delta H = E_A E_B + C_A C_B$$

to estimate enthalpies of reaction between Lewis acids and bases [146]. In this equation $A$ and $B$ indicate the acid and base while $E$ and $C$ are two empirically derived parameters assigned to each. Values for $E$ and $C$ are initially obtained from calorimetric measurements and are normalized to $E_A = E_B = 1$ for iodine. The product of $E_A$ and $E_B$ is considered to be related to the electrostatic contribution to the bond and the product of $C_A$ and $C_B$ related to the contribution from covalent interactions. Calculated enthalpies of adduct formation for a range of molecular species agree with experimentally determined values to within ± 0.2 kcal per mole. Although the actual significance of the $E$ and $C$ parameters is not known, the method may prove useful in predicting and interpreting donor-acceptor reactions.

In another approach to the estimation of solvent polarities the effect of a solvent on the absorbance maximum in the visible-ultraviolet region of the charge-transfer band of a salt such as 1-ethyl-4-carbomethoxy pyridinium iodide is measured [147]. A shift of the maximum to shorter wavelengths occurs as solvent polarity increases. The wavelength, expressed in kcal, is called the $Z$ value of the solvent. This method provides a simple and rapid measure of solvent polarity at the molecular level.

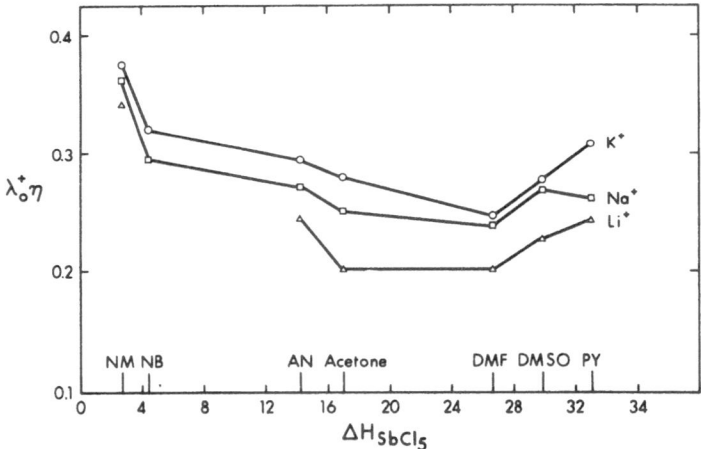

Fig. 10. Limiting single ion mobility-viscosity product for alkali metal ions *vs.* solvent Lewis basicity (as measured by enthalpy of reaction with $SbCl_5$)

A third method of estimating solvent basicity is provided by the donor number concept [148]. The donor number of a solvent is the enthalpy of reaction, measured in kcal per mole, between the solvent and a Lewis acid such as antimony (V) chloride. (Other Lewis acids, such as iodine or trimethyltin chloride, may be used, but the scale most often reported is that for $SbCl_5$.) Available values for the $SbCl_5$ donor number have been included in Table 1. Plots of the Walden product *versus* solvent basicity ($\Delta H_{SbCl_5}$) for several solvents are shown for lithium, sodium, and potassium ions in Fig. 10 and for the tetraalkylammonium ions in Fig. 11. A few rough correlations may be pointed out. For the alkali metals the general trend is for mobility to decrease as the solvent basicity increases. Dimethylsulfoxide and pyridine do not follow the trend; the anomalously high mobilities found in dimethylsulfoxide may be the result of solvent structure effects, but the results in pyridine are not readily explained. The tetraalkylammonium ions show little general trend, as expected. Again dimethylsulfoxide and pyridine do not fit the pattern set by the other solvents.

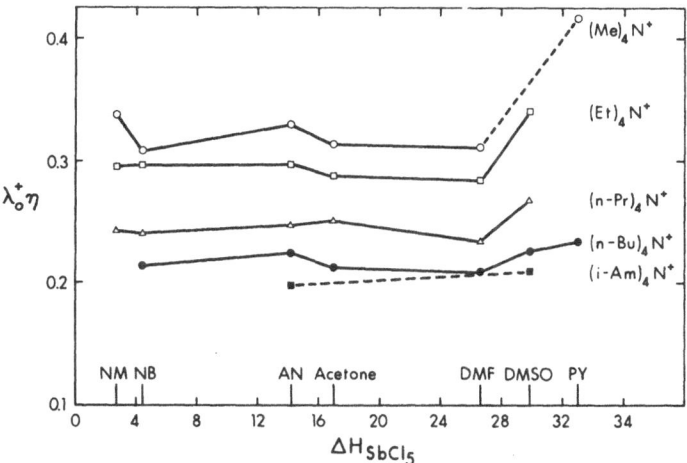

Fig. 11. Limiting single ion mobility-viscosity product for tetraalkylammonium ions *vs.* solvent Lewis basicity

There is no correlation between $Z$ values and donor numbers, as they measure different properties. Other methods of estimating solvent polarities have been discussed by Kosower, who has tabulated values for many of them [147].

## C. Conductance Studies in Solvent Mixtures

A considerable volume of literature has accumulated on conductance measurements in mixtures of solvents. Ion mobilities and association constants have been measured over a range of bulk dielectric constants with the aim of correlating bulk solvent properties with mobilities, ion association, and ion size parameters. An example of a widely used solvent mixture is water and 1,4-dioxane, which are miscible over all concentrations, providing a dielectric constant range of 2 to 78. The data obtained in systems containing two or more solvents must be treated with circumspection, as one solvent may interact more strongly with a given species present in solution than the other, and the resulting environment in the vicinity of the ion may differ greatly from the bulk composition of the solution mixture. Further, the solvent components may not be distributed homogeneously throughout the mixture, but may be present in clusters. Thus a hydrogen-bonding solvent may tend to form a cluster of molecules from which the second solvent is excluded by its lack of hydrogen-bonding capability. Because of the complexity and the difficulties associated with their interpretation, solvent mixtures are not included in this review.

# IV. Summary

Conductance information on a wide variety of solvents and solutes is now available, though many gaps exist. Refinements in cell construction, along with attention to solvent and solute purities, make possible the collection of experimental data of high precision. This data when treated by the conductance equations of Fuoss yield precise limiting equivalent conductances for 1 : 1 electrolytes. In conjunction with transport number measurements, single ion mobilities are obtained that give insight into ion solvation. Unfortunately, accurate transport numbers have been measured in few solvents, and ion mobilities in the remainder must be based on a split in the limiting conductance of a salt composed of a large symmetrical cation and anion. Mobilities based on such splits differ from true mobilities by 1 to 2 % even in ideal cases.

Small ion association constants (down to 5 or so) can be determined with relatively high precision by conductance techniques. The magnitude of these constants is related to many of the factors that determine mobilities.

Association and mobilities are related in a complex way to the bulk properties of the solvent and solute. These properties include the charge density and distribution on the ions and the Lewis base properties, the strength and nature of the solvent molecule dipole, the hydrogen-bonding capability, and the intermolecular structure of the solvent. Some correlations can be made on the basis of mobility and association trends in series such as the halides and alkali metals within a single solvent; others can be drawn between solvents for a given ion. It appears that conductance measurements provide a clear measure of the sum of ion-solvent interactions, but that other techniques must be used in conjunction with conductance if assessments of individual contributions from specific factors are to be made.

# V. Acknowledgments

The authors wish to thank Barbara Butterwick and Glenn Johanson for assistance in preparation of the manuscript.

# VI. References

[1] Elias, L., Schiff, H. I.: J. Phys. Chem. *60*, 595 (1956). Also see comment by H. I. Schiff in Ref. 21.

[2] Robinson, R. A., Stokes, R. H.: Electrolyte Solutions, 2nd ed. London: Butterworth and Co. Ltd. 1959.

[3] Shedlovsky, T.: Conductometry, in: Physical Methods of Organic Chemistry (A. Weissberger, ed.), 3rd ed., Chap. 45. New York: Interscience Publishers 1960.

[4] Jones, G., Josephs, R. C.: J. Am. Chem. Soc. *50*, 1049 (1928).

[5] Shedlovsky, T.: J. Am. Chem. Soc. *52*, 1793 (1930).

[6] Dike, P. H.: Rev. Sci. Instr. *2*, 379 (1931).

[7] Feates, F. S., Ives, D. J. G., Pryor, J. H.: J. Electrochem. Soc. *103*, 580 (1956).

[8] Hawes, J. L., Kay, R. L.: J. Phys. Chem. *69*, 2420 (1965).

[9] Prue, J. E., Sherrington, P. J.: Trans. Faraday Soc. *57*, 1795 (1961).

[10] Yeager, H. L., Kratochvil, B.: J. Phys. Chem. *73*, 1963 (1969).

[11] Barthel, J.: Angew. Chem. Intern. Ed. *7*, 260 (1968).

[12] Lind, J. E., Fuoss, R. M.: J. Phys. Chem. *65*, 999 (1961).

[13] Janz, G. J., Danyluk, S. S.: J. Am. Chem. Soc. *81*, 3854 (1959).

[14] Della Monica, M., Lamanna, U.: J. Phys. Chem. *72*, 4329 (1969).

[15] Evans, D. F., Zawoyski, C., Kay, R. L.: J. Phys. Chem. *69*, 3878 (1965).

[16] Kay, R. L., Hales, B. J., Cunningham, G. P.: J. Phys. Chem. *71*, 3925 (1967).

[17] Daggett, H. M., Bair, E. J., Kraus, C. A.: J. Am. Chem. Soc. *73*, 799 (1951).

[18] Ames, D. P., Sears, P. G.: J. Phys. Chem. *59*, 16 (1955).

[19] Jones, G., Bollinger, G. M.: J. Am. Chem. Soc. *53*, 411 (1931).

[20] Shedlovsky, T.: J. Am. Chem. Soc. *54*, 1411 (1932).

[21] Nichol, J. C., Fuoss, R. M.: J. Phys. Chem. *58*, 696 (1954).

[22] Ives, D. J. G., Pryor, J. H.: J. Chem. Soc. *1955*, 2104.

[23] Parker, H. C.: J. Am. Chem. Soc. *45*, 1366, 2017 (1923).

[24] Mukerjee, P., Mysels, K. J., Dulin, C. I.: J. Phys. Chem. *62*, 1390 (1958).

[25] Mysels, E. K., Scholten, P. C., Mysels, K. J.: J. Phys. Chem. *74*, 1147 (1970).

[26] Sears, P. G., Wilhoit, E. D., Dawson, L. R.: J. Chem. Phys. *23*, 1274 (1955).

[27] Cunningham, G. P., Evans, D. F., Kay, R. L.: J. Phys. Chem. *70*, 3998 (1966).

[28] Tomkins, R. P. T., Janz, G. J., Andalaft, E.: J. Electrochem. Soc. *117*, 906 (1970).

[29] Hoover, T. B.: J. Phys. Chem. *74*, 2667 (1970).

[30] Grahame, D. C.: J. Electrochem. Soc. *99*, 370C (1952).

[31] Arrington, D. E., Griswold, E.: J. Phys. Chem. *74*, 123 (1970).

[32] Jones, G., Bradshaw, B. C.: J. Am. Chem. Soc. *55*, 1780 (1933).

[33] Lind, J. E., Zwolenik, J. J., Fuoss, R. M.: J. Am. Chem. Soc. *81*, 1557 (1959).

[34] Fuoss, R. M., Hsia, K. L.: Proc. Nat. Acad. Sci. U. S. *57*, 1550 (1967).

[35] Conway, B. E.: „Some Aspects of the Thermodynamic and Transport Behavior of Electrolytes, in: Physical Chemistry, An Advanced Treatise (H. Eyring, ed.), Vol. 9A, Chap. 1. New York: Academic Press 1970.

[36] Fuoss, R. M., Accascina, F.: Electrolytic Conductance. New York: Interscience 1959.

[37] Pitts, E.: Proc. Roy. Soc. *217A*, 43 (1953).

[38] Conway, B. E., Barradas, R. G. (eds.): Chemical Physics of Ionic Solutions, New York: Wiley 1966.

[39] Fernandez-Prini, R., Prue, J. E.: Z. Physik. Chem. *228*, 373 (1965).

[40] Fuoss, R. M., Onsager, L., Skinner, J. F.: J. Phys. Chem. *69*, 2581 (1965).

[41] Jones, G., Dole, N.: J. Am. Chem. Soc. *51*, 2950 (1929).

42) Fuoss, R. M.: J. Am. Chem. Soc. 79, 3301 (1957).
43) Kay, R. L.: J. Am. Chem. Soc. 82, 2099 (1960).
44) Spiro, M.: Determination of Transference Numbers, in: Physical Methods of Organic Chemistry (A. Weissberger, ed.) 3rd ed., Chap. 46. New York: Interscience 1960.
45) Milios, P., Newman, J.: J. Phys. Chem. 73, 298 (1969).
46) Kay, R. L., Vidulich, G. A., Fratiello, A.: Chem. Instr. 1, 361 (1969).
47) Davies, J. A., Kay, R. L., Gordon, A. R.: J. Chem. Phys. 19, 749 (1951).
48) Graham, J. R., Gordon, A. R.: J. Am. Chem. Soc. 79, 2350 (1957).
49) Blum, S., Schiff, H. I.: J. Phys. Chem. 67, 1220 (1963).
50) Springer, C. H., Coetzee, J. F., Kay, R. L.: J. Phys. Chem. 73, 471 (1969).
51) Kraus, C. A.: Ann. N. Y. Acad. Sci. 51, 789 (1949).
52) Fuoss, R. M., Hirsch, E.: J. Am. Chem. Soc. 82, 1013 (1960).
53) Coplan, M. A., Fuoss, R. M.: J. Phys. Chem. 68, 1181 (1964).
54) Coetzee, J. F., Cunningham, G. P.: J. Am. Chem. Soc. 87, 2529 (1965).
55) Janz, G., Ahmad, I., Venkatasetty, H.: J. Phys. Chem. 68, 889 (1964).
56) Thomas, A. B., Rochow, E. G.: J. Am. Chem. Soc. 79, 1843 (1957).
57) Sears, P. G., Wolford, R. K., Dawson, L. R.: J. Electrochem. Soc. 103, 633 (1956).
58) Savedoff, L. G.: J. Am. Chem. Soc. 88, 664 (1966).
59) D'Aprano, A., Triolo, R.: J. Phys. Chem. 71, 3474 (1967).
60) Accascina, F., Schiavo, S.: Ann. Chim (Italy) 43, 695 (1953); CA 48, 6205.
61) – Pistoia, G., Schiavo, S.: Ric. Sci. Rend. 7, 560 (1966).
62) Martin, A. R.: J. Chem. Soc. 1928, 3270.
63) Reynolds, M. B., Kraus, C. A.: J. Am. Chem. Soc. 70, 1709 (1948).
64) Senne, J. K., Kratochvil, B.: Anal. Chem. 43, 79 (1971).
65) Lester, G. R., Gover, T. A., Sears, P. G.: J. Phys. Chem. 60, 1076 (1956).
66) Dawson, L. R., Wilhoit, E. D., Holmes, R. R., Sears, P. G.: J. Am. Chem. Soc. 79, 3004 (1957).
67) Walden, P., Ulich, H., Busch, G.: Z. Phys. Chem. 123, 429 (1926).
68) Tomkins, R. P. T., Andalaft, E., Janz, G. J.: Trans. Faraday Soc. 65, 1906 (1969).
69) Sears, P. G., Wilhoit, E. D., Dawson, L. R.: J. Phys. Chem. 60, 169 (1956).
70) Yeager, H. L., Kratochvil, B.: J. Phys. Chem. 74, 963 (1970).
71) Conti, F., Pistoia, G.: J. Phys. Chem. 72, 2245 (1968).
72) Dippy, J. F. J., Hughes, S. R. C.: J. Chem. Soc. 1954, 953.
73) Janz, G. J., Marcinkowsky, A. E., Ahmad, I.: Electrochim. Acta 9, 1687 (1964).
74) Pistoia, G., Pecci, G.: J. Phys. Chem. 74, 1450 (1970).
75) Janz, G. J., Marcinkowsky, A. E., Ahmad, I.: J. Electrochem. Soc. 112, 104 (1965).
76) Bolzan, J. A., Arvia, A. J.: Electrochim. Acta 15, 39 (1970).
77) Ogston, A. G.: Trans. Faraday Soc. 32, 1679 (1936).
78) Parfill, G. D., Smith, A. L.: Trans. Faraday Soc. 59, 257 (1963).
79) Tewari, P. H., Johari, G. P.: J. Phys. Chem. 67, 512 (1963).
80) Kempa, R. F., Lee, W. H.: J. Chem. Soc. 1961, 100.
81) Atlani, C., Justice, J. C., Quintin, M., Dubois, J. E.: J. Chim. Phys. Physicochim. Biol. 66, 180 (1969).
82) Sears, P. G., Lester, G. R., Dawson, L. R.: J. Phys. Chem. 60, 1433 (1956).
83) Dawson, L. R., Wilhoit, E. D., Sears, P. G.: J. Am. Chem. Soc. 79, 5906 (1957).
84) Coplan, M. A., Fuoss, R. M.: J. Phys. Chem. 68, 1177 (1964).
85) Brusset, H., Kikindai, M.: Bull. Soc. Chim. Fr. 1962, 1150.
86) Crisp, S., Hughes, S. R. C., Price, D. H.: J. Chem. Soc. (A) 1968, 603.
87) French, C. M., Glover, K. H.: Trans. Faraday Soc. 51, 1427 (1955).
88) – Trans. Faraday Soc. 51, 1418 (1955).
89) Wright, C. P., Murray-Rust, D. M., Hartley, H.: J. Chem. Soc. 1931, 199.

90) Witschonke, C. R., Kraus, C. A.: J. Am. Chem. Soc. *69*, 2472 (1947).
91) Phillip, J. C., Oakley, H. B.: J. Chem. Soc. *125*, 1189 (1924).
92) Murray-Rust, D. M., Hadow, H. J., Hartley, H.: J. Chem. Soc. *1931*, 215.
93) Della Monica, M., Lamanna, U., Jannelli, L.: Gazz. Chim. Ital. *97*, 367 (1967).
94) Mukherjee, L. M., Boden, D. P.: J. Phys. Chem. *73*, 3965 (1969).
95) Fernandez-Prini, R., Prue, J. E.: Trans. Faraday Soc. *62*, 1257 (1966).
96) Burgess, D. S., Kraus, C. A.: J. Am. Chem. Soc. *70*, 706 (1948).
97) Bhattacharyya, D. N., Lee, C. L., Smid, J., Szwarc, M.: J. Phys. Chem. *69*, 608 (1965).
98) Della Monica, M., Lamanna, U.: Gazz. Chim. Ital. *98*, 256 (1968).
99) Harris, W. S., Thesis, Ph. D.: University of California, Berkley, California 1958.
100) Hoover, T. B.: J. Phys. Chem. *68*, 876 (1964).
101) Mukherjee, L. M., Boden, D. P., Lindauer, R.: J. Phys. Chem. *74*, 1942 (1970).
102) Luder, W. F., Kraus, C. A.: J. Am. Chem. Soc. *69*, 2481 (1947).
103) Berns, D. S., Fuoss, R. M.: J. Am. Chem. Soc. *82*, 5585 (1960).
104) Eliassaf, J., Fuoss, R. M., Lind, J. E.,Jr.: J. Phys. Chem. *67*, 1941 (1963).
105) Popov, A. I., Skelly, N. E.: J. Am. Chem. Soc. *76*, 5309 (1954).
106) Evans, D. F., Gardam, P.: J. Phys. Chem. *73*, 158 (1969).
107) Sears, P. G., Wilhoit, E. D., Dawson, L. R.: J. Phys. Chem. *59*, 373 (1955).
108) Adams, W. A., Laidler, K. J.: Can. J. Chem. *46*, 1977 (1968).
109) McDowell, M. J., Kraus, C. A.: J. Am. Chem. Soc. *73*, 3293 (1951).
110) Ahmed, I. Y., Schmulbach, C. K.: Inorg. Chem. *8*, 1411 (1969).
111) — Schmulbach, C. D.: J. Phys. Chem. *71*, 2358 (1967).
112) Kortüm, G., Hokhale, S. D., Wilski, H.: Z. Phys. Chem. (Frankfurt/Main) *4*, 86 (1955).
113) French, C. M., Muggleton, D. F.: J. Chem. Soc. *1957*, 2131.
114) Hirsch, E., Fuoss, R. M.: J. Am. Chem. Soc. *82*, 1018 (1960).
115) Treiner, C., Fuoss, R. M.: Z. Phys. Chem. (Leipzig) *228*, 343 (1965).
116) Mead, T. H., Hughes, O. L., Hartley, H.: J. Chem. Soc. *1933*, 207.
117) Evans, D. F., Gardam, P.: J. Phys. Chem. *72*, 3281 (1968).
118) Kay, R. L., Zawoyski, C., Evans, D. F.: J. Phys. Chem. *69*, 4208 (1965).
119) Conti, F., Delogu, P., Pistoia, G.: J. Phys. Chem. *72*, 1396 (1968).
120) Barak, M., Hartley, H.: Z. Phys. Chem. (Leipzig) *A165*, 273 (1933).
121) Whorton, R., Amis, E. S.: Z. Phys. Chem. (Frankfurt/Main) *17*, 300 (1958).
122) Evers, E. C., Knox, A. G.: J. Am. Chem. Soc. *73*, 1739 (1951).
123) Accascina, F., Petrucci, S.: Ric. Sci. Suppl. *29*, 1383 (1959); Accascina, F., Antonucci, L.: Ric. Sci. Suppl. *29*, 1391 (1959); through Ref. 73.
124) Kay, R. L., Evans, D. F., Cunningham, G. P.: J. Phys. Chem. *73*, 3322 (1969).
125) — Blum, S. C., Schiff, H. I.: J. Phys. Chem. *67*, 1223 (1963).
126) Hughes, S. R. C., Price, D. H.: J. Chem. Soc. (A) *1967*, 1093.
127) Singh, R. D., Rastogi, P. P., Gopal, R.: Can. J. Chem. *46*, 3525 (1968).
128) Taylor, E. G., Kraus, C. A.: J. Am. Chem. Soc. *69*, 1731 (1947).
129) Grover, T. A., Sears, P. G.: J. Phys. Chem. *60*, 330 (1956).
130) Hughes, S. R. C., Price, D. H.: J. Chem. Soc. (A) *1968*, 1464.
131) Longo, F. R., Kerstetter, J. D., Kumosinski, T. F., Evers, E. C.: J. Phys. Chem. *70*, 431 (1966).
132) Banewicz, J. J., Maguire, J. A., Su-Shih, P.: J. Phys. Chem. *72*, 1960 (1968).
133) Brown, A. B., Fuoss, R. M.: J. Phys. Chem. *64*, 1341 (1960).
134) Kunze, R. W., Fuoss, R. M.: J. Phys. Chem. *67*, 385 (1963).
135) Venkatasetty, H. V., Brown, G. H.: J. Phys. Chem. *66*, 2075 (1962).
136) Morel, J. P.: Bull. Soc. Chim. Fr. *1966*, 1405.
137) Dunnett, J. S., Gasser, R. P. H.: Trans. Faraday Soc. *61*, 922 (1965).
138) Notley, J. M., Spiro, M.: J. Phys. Chem. *70*, 1502 (1966).

B. Kratochvil and H. L. Yeager

139)   Kay, R. L., Hawes, J. L.: J. Phys. Chem. *69*, 2787 (1965).
140)   Taylor, E. G., Kraus, C. A.: J. Am. Chem. Soc. *69*, 1731 (1947).
141)   Della Monica, M., Lamanna, U., Senatore, L.: J. Phys. Chem. *72*, 2124 (1968).
142)   Sears, P. G., Caruso, J. A., Popov, A. I.: J. Phys. Chem. *71*, 905 (1967).
143)   Dyke, M. D., Sears, P. G., Popov, A. I.: J. Phys. Chem. *71*, 4140 (1967).
144)   Kratochvil, B., Lorah, E., Garber, C.: Anal. Chem. *41*, 1793 (1969).
145)   Janz, G. J., Tait, M. J., Meier, J.: J. Phys. Chem. *71*, 963 (1967).
146)   Drago, R. S., Wayland, B. B.: J. Am. Chem. Soc. *87*, 3571 (1965).
147)   Kosower, E. M.: An Introduction to Physical Organic Chemistry, Section 2.6. New York: Wiley 1968.
148)   Gutmann, V.: Coordination Chemistry in Nonaqueous Solutions, Chap. 2. New York: Springer 1968.
149)   Zwanzig, R.: J. Chem. Phys. *52*, 3625 (1970).
150)   R. Fernandez-Prini, R., Atkinson, G.: J. Phys. Chem. *75*, 239 (1971).

Received January 22, 1971

# Ionic and Redox Equilibria in Donor Solvents

**Prof. Dr. Viktor Gutmann**

Institut für anorganische Chemie der Technischen Hochschule Wien

## Contents

# 1. Introduction

## 1.1. The Electrostatic Description of Ionization in Solution

The theory of complete electrolytic dissociation at infinite dilution was developed by Debye and Hückel (1923) and has been further extended by Onsager and later by Fuoss and Krauss.

The electrostatic description of ion formation in solution is satisfactory as long as ionic compounds are dissolved in a solvent [1, 2]. The energy required to dissolve an ionic compound is furnished by the interaction of the ions with the solvent molecules (Fig. 1): the ions are surrounded by a number of solvent molecules, and thus are "solvated".

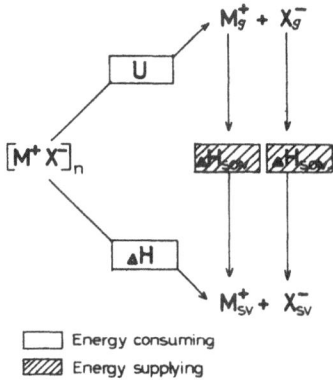

☐ Energy consuming

▨ Energy supplying

Fig. 1. Born-Haber cycle for the formation of solvated ions from an ionic crystal $[M^+X^-]_n$. U lattice energy, $\Delta H_{solv.}$ enthalpy of ion solvation

According to the electrostatic model the solvation is due to electrostatic interaction between the charged ions and the dipolar solvent molecules. Thus the solvating and ionizing properties of a solvent are considered as being due primarily to the *dipole moment* of the solvent molecules. Thus, ionic compounds such as sodium chloride are insoluble in non-polar solvents such as carbon tetrachloride. Actually, rather than the dipole moment the *"field action of the dipoles"* should be considered. This approach might explain why acetonitrile ($\mu = 3.2$) is poor in its ionizing properties compared to water ($\mu = 1.84$). However, no numerical values are available for this quantity.

The *equilibrium constant* for ionic dissociation is a measure of the extent of the separation of the ions in the system under consideration. It is unfortunate that this quantity is frequently termed "ionization constant". This definitely arises because the electrostatic theory does not distinguish between the process of ionization and that of ionic dissociation.

The value of the ionic dissociation constant for a given equilibrium is strongly influenced by the dielectric constant ($\epsilon$) of the medium: according to Coulomb's law the force acting between the charged particles $e_1{}^+$ and $e_2{}^-$ increases with decreasing $\epsilon$ values:

$$K = \frac{1}{\epsilon} \cdot \frac{e_1^+ \cdot e_2^-}{r^2}$$

Thus, in a medium of low dielectric constant the ions will undergo ion association. *Associated ions,* such as ion pairs of 1 : 1 electrolytes will not contribute to the conductivity of the solution at low field strengths. Furthermore, Coulomb's law explains why ions of equal charge but of different size are associated to a different degree: in a medium of given dielectric constant a compound consisting of big ions is more dissociated than one with small ions: cesium hydroxide is a stronger base than potassium hydroxide. On the other hand, various halides of the alkali metal ions do not obey this law [2].

No explanation is provided by the electrostatic model for the different behavior of ions of equal size and equal charge: the enthalpy of hydration is larger for $Hg^{2+}$ than for $Sr^{2+}$ and the hydrated $Sr^{2+}$ is an extremely weak acid whereas the hydrated $Hg^{2+}$ is a much stronger acid.

Furthermore, the electrostatic theory cannot explain why a given material may behave very differently in two different media of the same dielectric constant: water as well as concentrated sulfuric acid have a dielectric constant of about 80. Perchloric acid is completely dissociated in water but is a nonelectrolyte in sulfuric acid, whereas triphenyl carbinol is completely dissociated in sulfuric acid and non-dissociated in water.

Thus the ionization of covalent compounds cannot be interpreted by the electrostatic theory unless covalent interactions between solute and solvent are considered. It is the purpose of the present discussion to reveal the role of donor-acceptor interactions for all solution processes.

## 1.2. Donor and Acceptor Functions of a Solvent

From the point of view of coordination chemistry a distinction has been suggested between coordinating and non-coordinating solvents [3]. On the basis of this approach coordinating solvents may be divided [4] into

1. *Donor solvents,* which have a tendency of react with electron pair acceptors, and

2. *Acceptor solvents,* which tend to react with electron pair donors.

Many characteristic features will be different for these two classes of solvents: the former tend to solvate primarily metal cations, while the latter prefer to solvate anions. Water and several other solvents are unique in acting both ways, as donor solvents (towards metal ions) and as acceptor solvents (towards anions); the donor functions are due to the electron pair available at the oxygen atom, whereas the acceptor functions are due to the tendency to form strong hydrogen bonds. It has been found useful to classify solvents according to their dominating function: water, alcohols, nitriles, ammonia, amines, compounds with C=O, P=O, S=O groups, and many others are considered as donor solvents D, while sulfuric acid, the hydrogen halides and certain covalent halides, such as $BF_3$, $AsCl_3$ are considered acceptor solvents A [4, 5].

| DMF | = dimethylformamide | DMA | = dimethylacetamide |
|------|------|------|------|
| AN | = acetonitrile | ES | = ethylenesulphite |
| DMSO | = dimethylsulphoxide | HMPA | = hexamethylphosphoricamide |
| PDC | = propanediol-1,2-carbonate | NM | = nitromethane |
| TMP | = trimethylphosphate | BN | = benzonitrile |

## 1.3 The Formation of EDA-molecular Compounds

A neutral donor is capable of reacting with various covalent compounds by nucleophilic attack at the electropositive partner:

$$D + \overset{\delta+}{M} - \overset{\delta-}{X} \rightleftharpoons D \rightarrow M - X$$

This event leads to the formation of an electron donor-acceptor (EDA) complex involving the formation of a coordinate link between D and M. The availability of the additional electron pair at M causes an increase in electron density at X due to further polarization of the M—X bond. It is apparent that the amount of polarization will depend on both the polarizability of the covalent bond as well as the extent of interaction between D and M. For a given substrate the latter will depend on the donor properties of the donor [8].

## 1.4. The Concept of Donicity (Donor Number)

Briegleb [6] suggested that the *"donor strength"* of a molecule is an absolute property of the molecule and represented by its ionization energy. This is defined as the energy required to remove an electron completely from a free gaseous molecule from its ground state. It is obvious, however, that in the course of the formation of a EDA-complex all electrons remain within the reaction product and hence the ionization energy may not be a truly representative entity. Indeed, data listed in Table 1 illustrate that ionization energies may be similar for donors of vastly different donor properties.

Table 1. *Ionization potentials I for various molecules*

| Compound | $I$ [eV] | Compound | $I$ [eV] |
|---|---|---|---|
| Trimethylamine | 7.8 | Amonia | 10.2 |
| Pyridine | 7.9 | Acetic acid | 10.4 |
| Xylene | 8.3 | Hexane | 10.4 |
| Benzene | 9.2 | Ethanol | 10.5 |
| Diethylether | 9.5 | Dichloromethane | 11.3 |
| Acetone | 9.7 | Chloroform | 11.4 |
| Carbon disulfide | 10.1 | Carbon tetrachloride | 11.5 |
| Ethylacetate | 10.1 | Water | 12.6 |

It has been suggested that the donor strength be defined relative to a reference acceptor, for which antimony pentachloride was arbitrarily selected [7], which forms 1 : 1 adducts, $D.SbCl_5$, with neutral donors. The negative $\Delta H_{D.SbCl_5}$-value in high dilution of 1,2-dichloroethane is considered to be a measure of the donor properties of D and is termed *"donicity"* or *"donor number"*, DN.

$$D + SbCl_5 \rightleftharpoons D.SbCl_5; \quad -\Delta H_{D.SbCl_5} \equiv DN$$

$$K_{D.SbCl_5} = \frac{a_{D.SbCl_5}}{a_D \cdot a_{SbCl_5}}$$

Since the $\log K_{D.SbCl_5}$ -values are proportional to the donicities (Fig. 2)

$$p \log D_{D.SbCl_5} + q = -\Delta H_{D.SbCl_5}$$

the $pK_{D.SbCl_5}$ values may also be used as a measure of the donor strength.

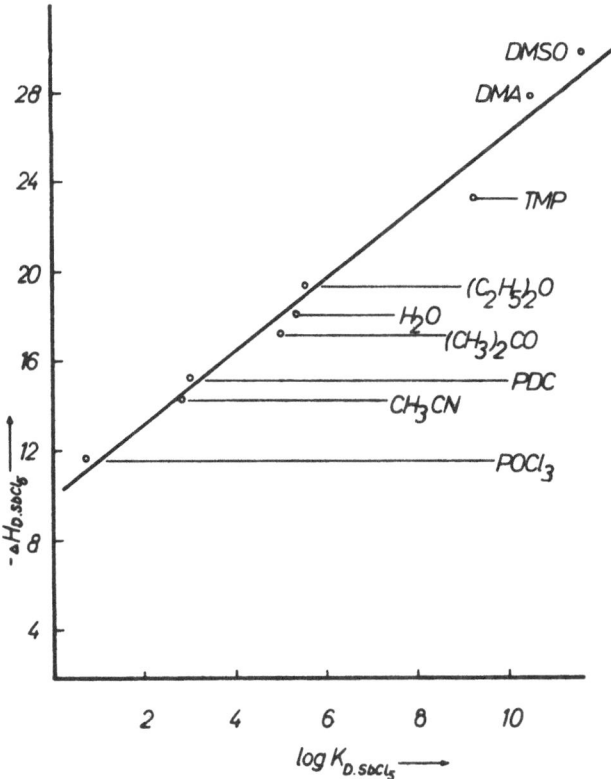

Fig. 2. Relationship between $-\Delta H_{D.SbCl_5}$ and log $K_{D.SbCl_5}$ for various donors D

The donicity represents the total enthalpy of interaction including the electrostatic constributions between D and $SbCl_5$ in high dilution of 1,2-dichloroethane [3, 7].

Table 2. *Donicities and dielectric constants*

| Solvent | $DN_{SbCl_5}$ | $\epsilon$ |
| --- | --- | --- |
| 1,2-Dichlorethane | – | 10.1 |
| Sulfurylchloride | 0.1 | 10.0 |
| Thionylchloride | 0.4 | 9.2 |
| Acetylchloride | 0.7 | 15.8 |
| Tetrachloroethylene carbonate | 0.8 | 9.2 |

Table 2 (continued)

| Solvent | $DN_{SbCl_5}$ | $\epsilon$ |
|---|---|---|
| Benzoylchloride | 2.3 | 23.0 |
| Nitromethane (NM) | 2.7 | 35.9 |
| Dichloroethylene carbonate (DEC) | 3.2 | 31.6 |
| Nitrobenzene (NB) | 4.4 | 34.8 |
| Acetic anhydride | 10.5 | 20.7 |
| Phosphorus oxychloride | 11.7 | 14.0 |
| Benzonitrile (BN) | 11.9 | 25.2 |
| Selenium oxychloride | 12.2 | 46.0 |
| Acetonitrile (AN) | 14.1 | 38.0 |
| Sulpholane | 14.8 | 42.0 |
| Propanediol-1,2-carbonate (PDC) | 15.1 | 69.0 |
| Benzylcyanide | 15.1 | 18.4 |
| Ethylenesulphite (ES) | 15.3 | 41.0 |
| iso-Butyronitrile | 15.4 | 20.4 |
| Propionitrile | 16.1 | 27.7 |
| Ethylenecarbonate (EC) | 16.4 | 89.1 |
| Phenylphosphonic difluoride | 16.4 | 27.9 |
| Methylacetate | 16.5 | 6.7 |
| n-Butyronitrile | 16.6 | 20.3 |
| Acetone | 17.0 | 20.7 |
| Ethylacetate | 17.1 | 6.0 |
| Water | 18.0 | 81.0 |
| Phenylphosphonic dichloride | 18.5 | 26.0 |
| Diethylether | 19.2 | 4.3 |
| Tetrahydrofurane (THF) | 20.0 | 7.6 |
| Diphenylphosphonic chloride | 22.4 | – |
| Trimethylphosphate (TMP) | 23.0 | 20.6 |
| Tributylphosphate (TBP) | 23.7 | 6.8 |
| Dimethylformamide (DMF) | 26.6 | 36.1 |
| N,N-Dimethylacetamide (DMA) | 27.8 | 38.9 |
| Dimethylsulphoxide (DMSO) | 29.8 | 45.0 |
| N,N-Diethylformamide | 30.9 | – |
| N,N-Diethylacetamide | 32.2 | – |
| Pyridine (py) | 33.1 | 12.3 |
| Hexamethylphosphoricamide (HMPA) | 38.8 | 30.0 |

The $\Delta H_{D.A}$-values towards A = $I_2$, $SbBr_3$, $(CH_3)_3SnCl$ and phenol are approximately proportional to the donicity $\Delta H_{D.SbCl_5}$ (Fig. 3). Since it is unlikely that this relationship will exist for any acceptor A, the donicity may be used only as an approximate expression of the donor strength of a molecule towards a given substrate, though it has been found to serve as a most useful guide for the interpretation or prediction of a number of interactions.

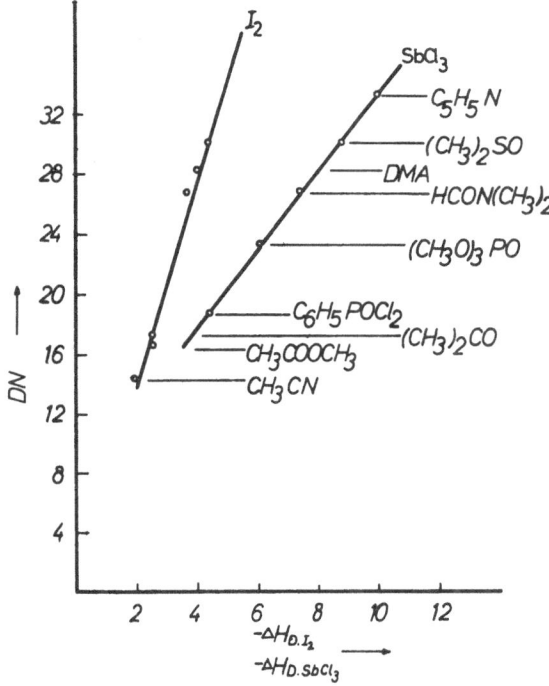

Fig. 3. Donicity and $\Delta H_{D \cdot A}$-values for A = $I_2$ or SbCl$_3$

## 1.5. Donicity, Polarizing Power and Ionization

Trifluoroiodomethane gives weak molecular complexes with various neutral donors. The $\Delta H$-values increase with increasing donicity of the donor molecule[8]. The interaction leads to the formation of $D_2 ICF_3$ compounds, in which the donor molecules appear to be bonded to the iodine atoms. The $^{19}$F NMR spectra demonstrate that the chemical shift is a function of both the molar ratio D:CF$_3$I (Fig. 4) and of the donicity of the donor molecule[8]. The plot of the δ-values vs. donicity reveals a linear relationship between the change in electron density within the CF$_3$-group and the donicity of the donor[8] (Fig. 5). It is apparent that complete transfer of the electron will result in a heterolytic fission of the bond, e.g. ionization of the covalent bond will occur, as will be discussed in more detail in the following sections.

67

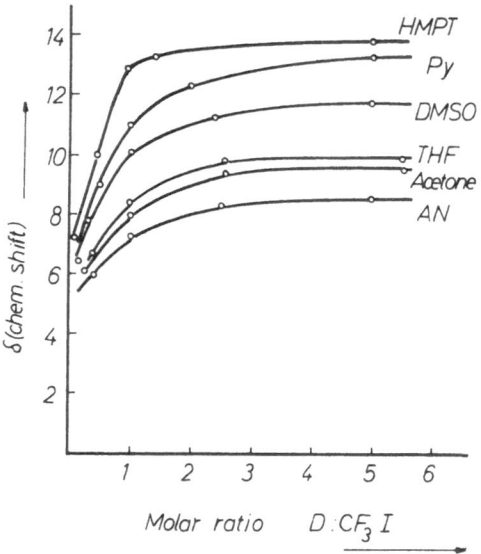

Fig. 4. $^{19}$F NMR results δ (ppm) of $CF_3I$ as a function of the molar ratio $D:CF_3I$ referred to $CCl_3F$ as external reference

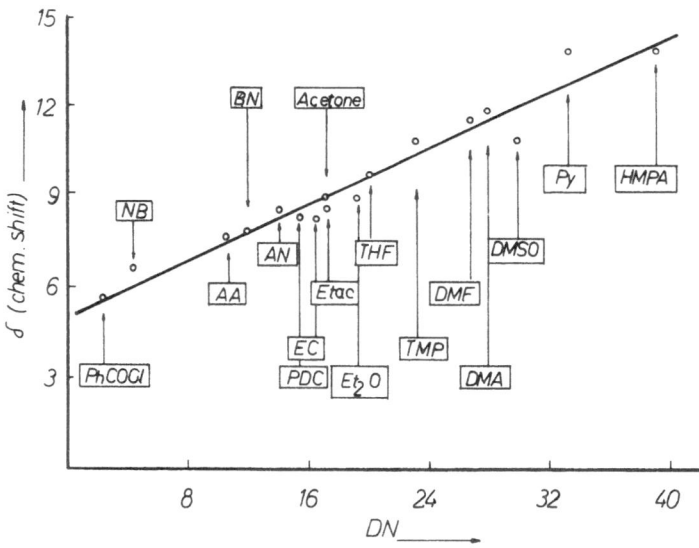

Fig. 5. $^{19}$F NMR results for the formation of $D-CF_3I$ complexes. δ (ppm) of $CF_3I$ at infinite dilution in the donor solvent D

# 2. The Ionization of Covalent Compounds

## 2.1. Introduction

*Heterolytic fission* of a covalent bond (ionization) will occur if the solvent-solute interaction permits the complete transfer of an electron from the more electropositive to the more electronegative partner of the bond [4, 5].

This event may be accomplished either by nucleophilic attack of the donor at M leading to a cation which is stabilized by coordination:

$$D + MX \rightleftharpoons D \rightarrow \overset{\frown}{M-X} \rightleftharpoons DM^+X^-$$

for example [9]:

$$2 D + AsI_3 \rightleftharpoons [D_2AsI_2]^+I^-$$

or by electrophilic attack of an acceptor molecule at X resulting in the formation of an anion which is stabilized by coordination:

$$M-X + A \rightleftharpoons \overset{\frown}{M-X} \rightarrow A \rightleftharpoons M^+XA^-$$

for example:

$$Ph_3CCl + A \rightleftharpoons [Ph_3C]^+[ClA]^-$$

The cited reactions may be considered as *ligand exchange reactions,* in which either a neutral donor D replaces an anion donor X at M or a neutral acceptor A replaces the cation at X.

The different behaviour of $HClO_4$ and $Ph_3COH$ in both water and sulfuric acid can now be interpreted: Perchloric acid, which has pronounced acceptor properties, reacts readily and completely with the donor solvent water and ionization occurs but there is no interaction with the acceptor solvent sulfuric acid and, hence, no ionization is observed in the latter solvent. Triphenylcarbinol, on the other hand, reacts with the acceptor sulfuric acid and not with the donor solvent water.

$$
\begin{aligned}
H_2O + HClO_4 &\rightleftharpoons & [H_3O]^+ + [ClO_4]^- \\
H_2SO_4 + HClO_4 &: & \text{no ionization} \\
Ph_3COH + H_2SO_4 &\rightleftharpoons & [Ph_3C]^+ + [HSO_4]^- + H_2O \\
Ph_3COH + H_2O &: & \text{no ionization}
\end{aligned}
$$

## 2.2. General Description of Ionization

Ionization of a covalent compound may be defined as the process leading to the formation of solvated ions independent of their presence as associated ions or as free entities (Fig. 6). In a medium of low dielectric constant the formation of associated ions is favored. It is therefore conceivable to consider the overall process of ionization as consisting of two steps, i.e., the *formation of associated ions* due to cation-coordination and anion-solvation and the *dissociation of the associated ions* in solution as a dielectric effect.

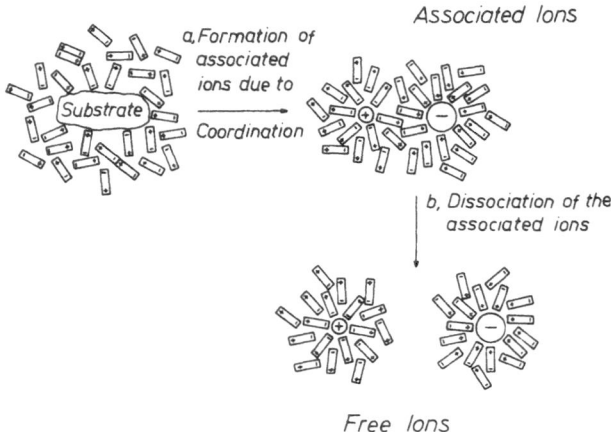

Fig. 6. Formation of associated ions and their dissociation

The consideration of a Born-Haber cycle shows that the energy-supplying terms for the ionization are apart from the electron affinity of X the solvation

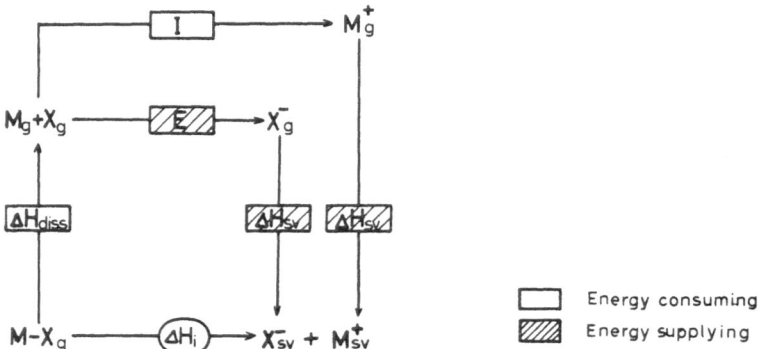

Fig. 7. Born-Haber cycle for the formation of ions by heterolytic fission of a covalent bond. I ionization potential of M, E electron affinity of X

energies of both cation and anion (Fig. 7). The total interactions between a donor solvent D and the substrate may be considered as due to both the *coordinating effect at the cation* [5, 10, 11] and the *solvation effect at the anion* [5, 12].

The equilibrium constant for the formation of associated ions may be termed *ionization constant* $K_{Ion}$

$$D + MX \rightleftharpoons [DM^+X^-]^\circ$$

$$K_{Ion} = \frac{a[DM^+X^-]^\circ}{a_D \cdot a_{MX}}$$

and

$$MX + A \rightleftharpoons [M^+XA^-]^\circ$$

$$K_{Ion} = \frac{a[M^+XA^-]^\circ}{a_{MX} \cdot a_A}$$

respectively.

For a given substrate $K_{Ion}$ is increased with increasing coordination between substrate and solvent.

The equilibrium constant for the formation of free ions from associated ions may be termed dissociation constant $K_{Diss}$, since the process represents the dissociation of associated ions. $K_{Diss}$ is a function of the dielectric constant of the medium. The reciprocal value $\frac{1}{K_{Diss}}$ is also termed association constant $K_{Ass}$ for the association of free ions.

$$[DM^+X^-]^\circ \rightleftharpoons DM^+ + X^-$$

$$\cdot K_{Diss} = \frac{a_{DM^+} \cdot a_{X^-}}{a[DM^+ \cdot X^-]^\circ} = \frac{1}{K_{Ass}}$$

and

$$[M^+AX^-]^\circ \rightleftharpoons M^+ + AX^-$$

$$K_{Diss} = \frac{a_{M^+} \cdot a_{AX^-}}{a[M^+AX^-]^\circ}$$

For the overall process of ionization both, the coordinating properties and the dielectric constant of the solvent must be considered.

## 2.3. The Donor Effect at the Cation

It has been pointed out, that the process of the formation of associated ions in a medium of $\epsilon = 1$ will be favored [11, 13]

  a) by increase in donicity of D,

b) by increase in polarizability of M–X, and

c) by decrease in bond energy of M–X.

The latter properties will also be influenced by the properties of the atoms M and X.

Acceptors may be considered as either *"hard" or "soft"*. Hard acceptors, such as the proton or alkali metal ions are hardly polarizable and tend to react preferentially with light donor atoms [14, 15]:

$$N \gg P > As > Sb$$
$$O \gg S > Se > Te$$
$$F \gg Cl > Br > I$$

Soft acceptors such as $Cu^+$, $Tl^+$, $Hg^{2+}$ are easily polarizable and tend to react with heavy donor atoms [14, 15]:

$$Sb < As < P \gg N$$
$$Te \sim Se \sim S \gg O$$
$$I > Br > Cl \gg F$$

Table 3. *Hard and soft metal ions*

| Hard | Border-line | Soft |
|------|-------------|------|
| $Li^+$, $Na^+$, $K^+$ | $Fe^{2+}$, $Co^{2+}$, $Ni^{2+}$, $Cu^{2+}$ | $Cu^+$, $Ag^+$, $Au^+$ |
| $Be^{2+}$, $Mg^{2+}$, $Ca^{2+}$, $Sr^{2+}$ | $Zn^{2+}$, $Pb^{2+}$ $Sn^{2+}$ | $Tl^+$, $Hg_2^{2+}$ |
| $Al^{3+}$, $Sc^{3+}$, $Ga^{3+}$ | $Sb^{3+}$, $B^{3+}$ | $Pd^{2+}$, $Cd^{2+}$, $Pt^{2+}$ |
| $In^{3+}$, $La^{3+}$, $Gd^{3+}$ | $Rh^{3+}$, $Ir^{3+}$ | $Hg^{2+}$ |
| $Cr^{3+}$, $Co^{3+}$, $Fe^{3+}$ | | $Pt^{4+}$, $Te^{4+}$ |
| $Si^{4+}$, $Ti^{4+}$, $Zr^{4+}$, $Th^{4+}$ | | $Tl^{3+}$ |
| $U^{4+}$, $Pu^{4+}$, $Ce^{3+}$, $Hf^{4+}$ | | $M^\circ$ (metal atoms) |
| $UO_2^{2+}$, $VO^{2+}$, $MoO^{3+}$ | | |

Hard-soft properties account for the differences in behavior of hydrogen halides and silver halides in water:

Although the H-F bond shows a higher polarization than the H-I bond, the latter is completely ionized in water since the H-I bond is weaker and more easily polarized than the latter.

$$H_2O + H-I \rightleftharpoons H_3O^+ + I^-; K_{Ion} > 1$$
$$H_2O + H-F \rightleftharpoons H_3O^+ + F^-; K_{Ion} \approx 10^{-4}$$

Likewise, the ionization of sodium halides increases in the order

$$NaF < NaCl < NaBr < NaI,$$

since the free enthalpies of dissolution $\Delta G_{sol}$, calculated from the free enthalpies of hydration $\Delta G_H$ and the free lattice enthalpies $\Delta G_L$

$$G_{sol} = \Delta G_H - \Delta G_L$$

decrease to negative values in the same order (Fig. 8).

The situation is, however, different for the halides of soft metal ions, such as silver: Here the solubility and hence the ionization in water decreases in the order

$$AgF \gg AgCl > AgBr > AgI$$

and indeed the free enthalpies of dissolution increase to positive values in the same order (Fig. 8). Addition of a donor which is stronger than water, such as ammonia, promotes both solubility and ionization but the order remains unchanged.

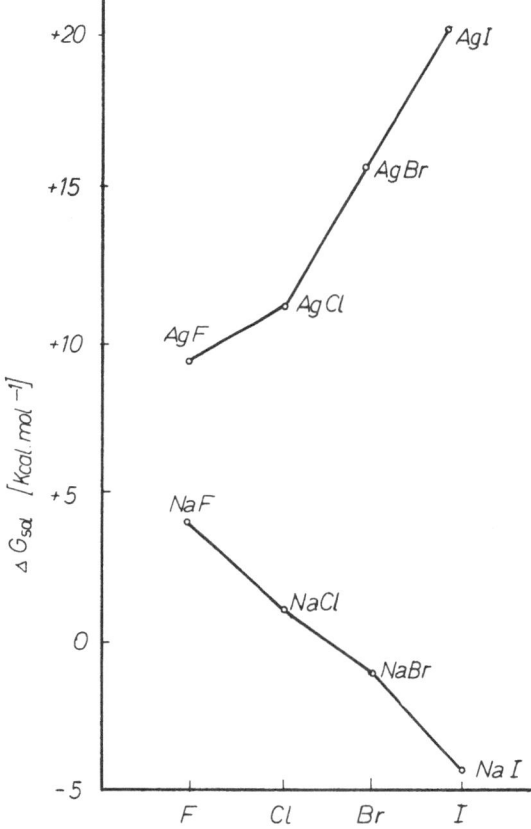

Fig. 8. Free enthalpies of solvation $\Delta G_{sol}$ for the halides of silver and sodium in water

73

Since the conductivity of a system is strongly influenced by the dielectric constant of the medium, it is not possible to draw conclusions from conductivity data of a given substrate in different donor solvents. Apart from other methods such as the NMR-technique, it has been found most useful to utilize the method of *conductometric titration* of the substrate with the donor in a medium of a reasonably high dielectric constant and which is as inert as possible as a coordinating agent. The physical properties of the medium remain essentially unchanged during the titration and the conductivity of the solution is related to the amount of associated ions present. If the contribution by solvation of the anions (see Sect. 2.4) is neglected, the conductivity is an indication of the donor effect at the cation. Typical media that can serve as inert solvents are

nitrobenzene ($DN = 4,4$, $\epsilon = 34.8$),
nitromethane ($DN = 2.7$, $\epsilon = 35.8$) and
dichloroethane ($DN \approx 0$, $\epsilon = 10.1$).

Covalent compounds, such as numerous halides of metals, give non-conducting solutions in these media.

The conductometric titrations of the non-conducting solution of trimethyltin iodide in nitrobenzene (at $c \approx 7.10^{-2}$) with different donors reveal that conducting solutions are formed. The molar conductivity at a given mole ratio D: $(CH_3)_3 SnI$ generally increases with increasing donicity of the donor [10] (Fig. 9), e.g.,

$$AN < THF < Ph_2 POCl < TBP < DMF < DMSO < HMPA.$$

In solution of pure HMPA, DMSO, or DMF, $(CH_3)_3 SnI$ is found to be completely ionized as 1 : 1 electrolyte. This observation leads to the conclusion that the equivalent conductivities are a measure of the relative ionizing power of the donor. *Thus we can say that the relative ionizing power of a donor solvent increases with an encrease in the donicity of the solvent molecules.*

Further evidence for this is provided by the *coupling constants J* ($^{119}Sn$-$CH_3$). The coupling constant of the nonionized substrate $J_f$ (in $CCl_4$) increases upon addition of a donor and the coupling constant of the ionized species $J_c$ is increased by increasing donicity of the donor.

As elaborated earlier for the behavior of hydrogen halides and alkali metal halides in water, the ionization of many other metal halides follows the same

Table 4. *Coupling constants J ($^{119}Sn$-$CH_3$)*

| Solvent | DN | $J_c [H_z]$ |
|---|---|---|
| HMPA | 38.8 | 72.0 |
| DMSO | 29.7 | 69.0 |
| DMF | 26.6 | 68.5 |
| Nitrobenzene | 4.4 | 59.0 |

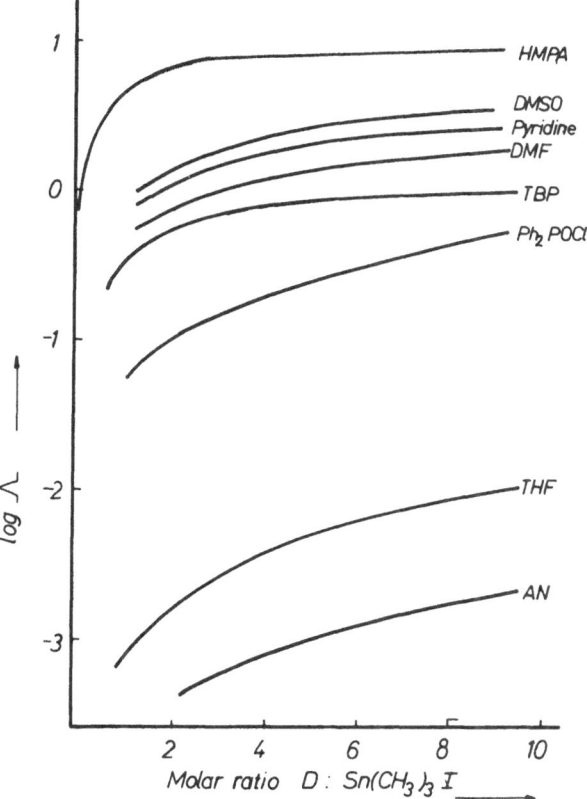

Fig. 9. Molar conductivities of $(CH_3)_3SnI$ in nitrobenzene as a function of the nature of D and the molar ratio

order in non-aqueous donor solvents. $SiF_4$ gives non-conducting adducts in liquid ammonia or pyridine, which remain non-ionic even in media of high dielectric constant. The adducts of $SiCl_4$ behave similarly and the claimed ionization reactions could not be confirmed [16, 17]. On the other hand, Si-I bonds are readily ionized by interaction with donor molecules [18 - 21].

$$4 \text{ py} + SiI_4 \rightleftharpoons [py_4SiI_2]^{2+} + 2 \text{ I}^-$$

## 2.4. The Donor Properties of Anions

Since the reactions may be described as ligand exchange reactions, the competition for coordination between neutral donor molecules and (ionized) anions is an important factor. Hence it is desirable to consider the donor strength of the (competitive) anions. It is, however, not possible to assign a donicity value to anions, since this is dependent not only on the nature of the acceptor, but also on the extent of solvation of the anions.

Table 5. *Free standard enthalpies* $\Delta G°$ *(VO) for the reactions VO (acac)$_2$AN + L $\rightleftharpoons$ VO(acac)$_2$L + AN in AN at 27°C*

| Ligand L | DN | $\Delta G°$ (VO) |
|----------|------|----------|
| I$^-$ | | 2.00 |
| PDC | 15.1 | + 1.75 |
| Acetone | 17.0 | + 1.30 |
| Br$^-$ | | + 0.65 |
| TMP | 23.0 | − 0.24 |
| DMF | 26.6. | − 0.63 |
| Cl$^-$ | | − 0.83 |
| DMSO | 29.8 | − 1.14 |
| Ph$_3$PO | 32.5 | − 2.11 |
| py | 33.1 | − 2.47 |
| HMPA | 38.8 | − 2.50 |
| F$^-$ | | − 2.69 |
| NCS$^-$ | | − 2.69 |
| CN$^-$ | | − 4.20 |
| N$_3$$^-$ | | − 4.20 |

In Table 5 the values for the free standard enthalpies for the reactions of neutral donors and anion donors with vanadyl acetylacetonate are listed. It can be seen that towards the reference molecule iodide ion is a somewhat weaker ligand than propanediol carbonate, whereas the bromide ion is between trimethylphosphate and acetone, and the chloride ion between DMF and DMSO [22]. The fluoride ion and the NCS$^-$-ion are stronger donors than all neutral donors but are somewhat weaker than the azide and the cyanide ion.

We shall now consider the reaction

$$[VO\,(acac)_2\,(DMSO)\,] + X^- \rightleftharpoons [VO\,(acac)_2 X]^- + DMSO$$

in acetonitrile, where $X^-$ is more strongly solvated than DMSO. The solvation of $X^-$ accounts for the decrease in donor strength of $X^-$ in AN as compared with that in the gas phase: the result being that the equilibrium between chloro complex and DMSO complex is more on the side of the DMSO-complex in AN than in the gas phase.

In Table 6 the differences of free enthalpies of solvation for several anion ligands in a donor solvent D and in AN are given. HMPA shows very weak solvation whereas water is a very strong solvating agent for anions. The free enthalpies of solvation of halide and pseudohalide ions are by 4 to 15 kcal/mol more negative than in aprotic donor solvents.

Table 6. *Differences* $\Delta G^{\circ}{}_{(D)}$-$\Delta G^{\circ}{}_{(AN)}$[kcal/mol] *of* $X^-$ *in D and in AN*

| $X^-$ | NM | AN | TMS | $H_2O$ | $CH_3OH$ | DMF | DMA | DMSO | HMPA |
|---|---|---|---|---|---|---|---|---|---|
| $I^-$ | 0.3 | 0 | 0 | −5.3 | −3.3 | 0.3 | 0.9 | −1.5 | |
| $Br^-$ | – | 0 | – | −8.6 | −5.7 | 1.0 | 2.4 | −0.8 | 4.0 |
| $Cl^-$ | −1.9 | 0 | −0.6 | −12.0 | −8.6 | 0.3 | 2.1 | −1.0 | 3.5 |
| $NCS^-$ | −0.3 | 0 | 0 | −5.2 | −3.5 | 0.1 | 0.8 | −1.6 | 1.1 |
| $N_3^-$ | −0.1 | 0 | 1.0 | −8.9 | −6.4 | 0.3 | 2.1 | −1.6 | 3.4 |

The strong solvation is due to hydrogen bridging. For this reason the ionizing properties of water are considerably stronger than is suggested by the consideration of its donicity.

Table 7. $\Delta G^{\circ}$, $\Delta H^{\circ}$ *and* $\Delta S^{\circ}$ *for the solvation of halide ions in DMSO and in water* [12]

| | Solvent | $I^-$ | $Br^-$ | $Cl^-$ |
|---|---|---|---|---|
| $\Delta G^{\circ}$ [kcal.$mol^{-1}$] | DMSO | −47.5 | −52.6 | −56.4 |
| | $H_2O$ | −51.3 | −60.4 | −67.4 |
| $\Delta H^{\circ}$ [kcal.$mol^{-1}$] | DMSO | −56.8 | −63.4 | −67.9 |
| | $H_2O$ | −54.3 | −64.7 | −72.8 |
| $\Delta S^{\circ}$ [cal.$grad^{-1}mol^{-1}$] | DMSO | −31.4 | −36.1 | −38.4 |
| | $H_2O$ | −10.1 | −14.4 | −18.2 |

Table 7 illustrates that the entropies of solvation are smaller in water than in DMSO. This can be attributed to the fact that the former has a more pronounced liquid structure.

The solvating properties of protonic solvents such as methanol, ethanol, or acetic acid, are intermediate between those of water and aprotic solvents. This factor accounts for many differences between water and the alcohols as solvents. For example, $CoBr_2$ is ionized in water, while in methanol tetrahedral $CoBr_2 \cdot (CH_3OH)_2$ is found [23].

Thus it is apparent that the ionizing properties of a solvent are determined by the sum of its coordinating properties, e.g. a factor must be added to its donocity which accounts for the solvation of the anion under consideration.

## 2.5. The Dielectric Contribution to Ionization

It has been pointed out, that the dielectric constant is an important quantity in determining the extent of dissociation of associated ions. The dissociation constants $K_{Diss}$ of quaternary ammonium salts may be considered as a qualitative indication of the dielectric properties of a solvent. For example, in a medium of $\epsilon \approx 10$ $K_{Diss} \approx 10^{-4}$ and in a medium of $\epsilon \approx 35$ $K_{Diss} \approx 10^{-2}$.

The dielectric properties of the solvent have also an influence on the ionization constant of an incompletely ionized substrate. By the process of ion dissociation the concentration of associated ions is decreased; this results because the latter are in equilibrium with non-ionized species and the ionization equilibrium will be restored by the formation of additional associated ions. In this manner the formation of ionized species is also favored by a high dielectric constant of the medium.

Table 8. *Extent of ionization of M—X at c $\approx 10^{-2}$ mol/l under assumed conditions*

| Solvent | $\epsilon$ | $K_{Diss}$ | $K_{Ion}$ | % Ionized |
|---------|-----|--------|-------|-----------|
| $D_1$ | 10 | $10^{-4}$ | 1 | 53 |
| $D_2$ | 35 | $10^{-2}$ | 1 | 75 |
| $D_3$ | 10 | $10^{-4}$ | 3 | 77 |

For the consideration of the ionizing properties of a solvent it is therefore imperative to consider both the coordinating and the dielectric properties [10]. An illustrative example is given in Table 8, in which three different donor solvents $D_1$, $D_2$ and $D_3$ are considered. $D_1$ and $D_3$ are assumed to have the same dielectric constant while $D_1$ and $D_2$ are assumed to have the same coordinating properties. The values for $\epsilon$ and $K_{Ion}$ have arbitrarily been selected. From $K_{Ion}$ and $K_{Diss}$ the sum of associated and dissociated ions can readily be calculated and the result shows that in $D_3$ the overall process of ionization is progressing further than in $D_2$, which has a much higher dielectric constant, but has somewhat smaller coordinating properties than $D_3$.

It is therefore apparent that dissociation constants *may only be compared in the same solvent*. Ammonia is a stronger donor than water, but liquid ammonia has a much lower dielectric constant than the latter. The acidity constant of hydrochloric acid in liquid ammonia is much lower than in water, in which it is completely ionized and completely dissociated, whereas the complete ionization in liquid ammonia is not followed by extensive ionic dissociation due to its low dielectric constant. On the other hand, the acidity constant of acetic acid is somewhat higher in liquid ammonia than in water since in the latter $K_{Ion}$ is much lower than in liquid ammonia, in which complete ionization is achieved.

*It is permissible to compare dissociation constants of different solutes in the same solvent, but it is meaningless to compare dissociation constants for the same compound in different solvents.*

## 2.6. A Few Further Examples

### 2.6.1. Organic Compounds

Numerous reactions in the field of organic chemistry are known to involve the intermediate formation of ions, though only few ionic equilibria are known. Amines, amides, alkoxides or halide ions are known to act as donors and to produce ionic species [24-26].

$$R_3N + RBr \rightleftharpoons [R_4N]^+Br^-$$
$$[C_6H_5NH]^- + F_2C{=}CF_2 \rightleftharpoons [C_6H_5NH{-}CF_2{-}CF_2]^-$$
$$[RO]^- + F_2C{=}CF_2 \rightleftharpoons [RO{-}CF_2{-}CF_2]^-$$
$$F^- + CF_2{=}CF{-}CF_3 \rightleftharpoons [CF_3{-}CF{-}CF_3]^-$$

The formation of coordinated carbonium ions is known to occur by the action of strong donor molecules,

$$Ph_3CCl + HMPA \rightleftharpoons [Ph_3C\,(HMPA)\,]^+Cl^-$$

but the ionization of methyl iodide by donor molecules has not been demonstrated. On the other hand, the ionization of $ICF_3$ occurs after nucleophilic attack of a donor at the iodine atom under the influence of light [8].

$$2\,D + ICF_3 \rightleftharpoons D_2ICF_3 \xrightarrow{h\nu} D_2I^+ + CF_3^-$$

Ionization of various organic compounds is known to be promoted by electrophilic attack of an acceptor, e.g. in the course of Friedel-Crafts reactions,

$$RCOCl + AlCl_3 \rightleftharpoons [RCO]^+[AlCl_4]^-$$

with formation carbonium ions,

$$(C_6H_5)_3CCl + HCl \rightleftharpoons [(C_6H_5)_3C]^+[HCl_2]^-$$
$$t{-}C_4H_9F + AsF_5 \rightleftharpoons [t{-}C_4H_9]^+[AsF_6]^-$$

or from chlorinated cyclic hydrocarbons [27]

$$\begin{bmatrix} Cl-C \\ \parallel \\ Cl-C \end{bmatrix} CCl_2 + FeCl_3 \; \rightleftharpoons \; \begin{bmatrix} Cl-C \\ | \\ Cl-C \end{bmatrix} C-Cl \; \end{bmatrix}^+ \quad [FeCl_4]^-$$

### 2.6.2. Organometallic Compounds

Benzylmagnesium chloride is ionized by HMPA [28]

$$C_6H_5CH_2MgCl + HMPA \; \rightleftharpoons \; [(HMPA)MgCl]^+[C_6H_5CH_2]^-$$

and likewise $\pi$-complexes are ionized by the action of strong neutral donors [29].

$$(\pi-C_4H_7)_2PdCl + 2\,(C_6H_5)_3P \; \rightleftharpoons \; [(\pi-C_4H_7)_2Pd(P(C_6H_5)_3)_2]^+Cl^-$$

The conductivities in acetonitrile are increased in the order

$$PPh_3 \; < \; PMe_2Ph \; < \; PEt_2Ph \; < \; PEt_3$$

as well as $I < Br < Cl$ in accordance with the reversed order of donor strength of halide ions towards soft metal ions.

Metal carbonyls may also be ionized by acceptor molecules, e.g. [30–33]

$$(CO)_3Mn\underset{Cl}{\bigcirc} \; + \; CO \; + \; AlCl_3 \; \rightleftharpoons \; [(Co)_4Mn\bigcirc \; ]^+ \; [AlCl_4]^-$$

$$M(CO)_n + X^- \; \rightleftharpoons \; [M(CO)_{n-1}X]^- \; +CO \text{ for } M = Cr, Mo, W, Ni, Re$$
$$\text{and } X = Cl, Br, I$$

Ferrocene, cobaltocene, and nickelocene, react with iodine (as well as with s-trinitrobenzene or picric acid [34]) to form ionic species:

$$2\,(Cy)_2Fe + 3\,I_2 \; \rightleftharpoons \; 2\,[(Cy)_2Fe]^+ + 2\,I_3^-$$

### 2.6.3. Compounds of Boron

Boron trichloride undergoes autocomplex formation by strong donors (see Sect. 4.4), but the iodide is easily ionized by interaction with pyridine [35]

$$2\,py + BI_3 \; \rightleftharpoons \; [(py)_2BI_2]^+ \; I^-$$

The adduct $(CH_3)_3NBH_2I$ is ionized even by much weaker donors, such as acetonitrile [36], but benzonitrile gives a non-ionic adduct.

$$(CH_3)_3NBH_2I + AN \rightleftharpoons [(CH_3)_3NBH_2(AN)]^+I^-$$

Another instructive example is the ionization of boron hydrides in the presence of strong donor molecules [37]

$$B_{10}H_{12}(AN)_2 + 2\,Et_3N \rightleftharpoons 2\,[Et_3NH]^+[B_{10}H_{10}]^- + 2\,AN$$
$$B_2H_6 + 2\,DMSO \rightleftharpoons [BH_2(DMSO)_2]^+[BH_4]^-$$

$[D_3BH]^{2+}$ cations are formed according to the equation [38]

$$(CH_3)_3NBHBr_2 + 3\,D \rightleftharpoons [D_3BH]^{2+} + 2\,Br^- + (CH_3)_3N$$

### 2.6.4. Compounds of Silicon

The ionization of silicon tetraiodide by pyridine has been mentioned above. Trichlorosilane is known to undergo ionization with amines in solution of acetonitrile [39]:

$$n-Pr_3N + HSiCl_3 \rightleftharpoons [n-Pr_3NH]^+[SiCl_3]^-$$

According to spectral evidence the complexes of silyl bromide and silyl iodide with strong donors are also ionic [40]:

$$[SiH_3(py)_2]^+\,Br^- \text{ and } [SiH_3(py)_2]^+I^-$$

They form conducting solutions in acetonitrile, while the corresponding chlorides and fluorides are non-conducting [40]. Triphenylsilyl iodide is ionized by 2,2'-bipyridyl in methylene chloride or acetonitrile [41]:

$$Ph_3SiI + bipy \rightleftharpoons [Ph_3Si(bipy)]^+I^-$$

Silyl cations are formed from silyltetracarbonyl cobalt, $H_3SiCo(CO)_4$, or silylpentacarbonyl manganese [42]:

$$H_3SiMn(CO)_5 + 2\,py \rightleftharpoons [H_3Si(py)_2]^+[Mn(CO)_5]^-$$

### 2.6.5. Metal Halides

The ionization of halides of hard or borderline metal ions is enhanced by decreased donor properties of the halide ion and by increased donor properties of the neutral donor. Thus cobalt (II) iodide and cobalt (II) bromide are completely ionized in dimethyl sulfoxide [43] (see Sect. 3.2.).

Nickel bromide is also completely ionized in DMSO

$$NiBr_2 + 6\,DMSO \rightleftharpoons [Ni(DMSO)_6]^{2+} + 2\,Br^-$$

but ionization in DMF is incomplete.

Vanadyl bromide is completely ionized in DMSO [45]:

$$VOBr_2 + n\,DMSO \rightleftharpoons [VO(DMSO)_n]^{2+} + 2\,Br^-$$

whereas vanadyl chloride is partly present as $[VOCl]^+$ cation [45].

$$VOCl_2 + m\,DMSO \rightleftharpoons [VOCl(DMSO)_m]^+ + Cl^-$$

In the presence of bibyridyl, titanium (III) bromide in AN is ionized to a greater extent than titanium (III) chloride [46].

$$2\,TiX_3 + 2\,bipy \rightleftharpoons [X_2Ti(bipy)_2]^+ [X_4Ti(bipy)]^-$$

Rhenium (III) halides appear to retain their cluster structure in donor solvents, although with strong donors ionization takes place [47].

$$Re_3Cl_9 + 6\,DMSO \rightleftharpoons [Re_3Cl_9(DMSO)_6]^{3+} + 3\,Cl^-$$

Rhenium (III) bromide is, as expected, more easily ionized, so that in DMSO the bisarsenate complex $Re_3Br_3(ASO_4)_2(DMSO)_3$ is readily produced [48].

The conductivities of thorium (III) and uranium (IV) halides in nitromethane are increased in the presence of a donor molecule from chloride to iodide and they also increase with increasing donicity of the neutral donor [49−53].

Iodides of soft metal ions such as mercury (II) are essentially un-ionized in dimethylsulfoxide [54]. This feature is due both to the strength of the Hg-I bond and the weakness of the Hg-DMSO bond which appears to occur through the sulfur atom of the DMSO molecule, as is known for the palladium (II) complex [55].

# 3. The Formation of Halo- and Pseudohalo-complexes

## 3.1. Donicity and Formation of Hexachloroantimonates

The ionization of a covalent halide in a donor solvent has been considered as a ligand exchange reaction or the replacement of $X^-$ by D at M. The formation of a halide complex from a solvate involves the replacement of D by $X^-$ at M.

It is apparent that for complex formation in solution the donor properties of the solvent should be as low as possible. Thus a good ionizing solvent will be a poor medium for complex formation since its molecules will compete for co-ordination with the ligands added to the solution.

While the reaction

$$D + SbCl_5 \rightleftharpoons D.SbCl_5$$

in dichloroethane is utilized to obtain values for the donor properties of the solvents, we shall now consider the reaction

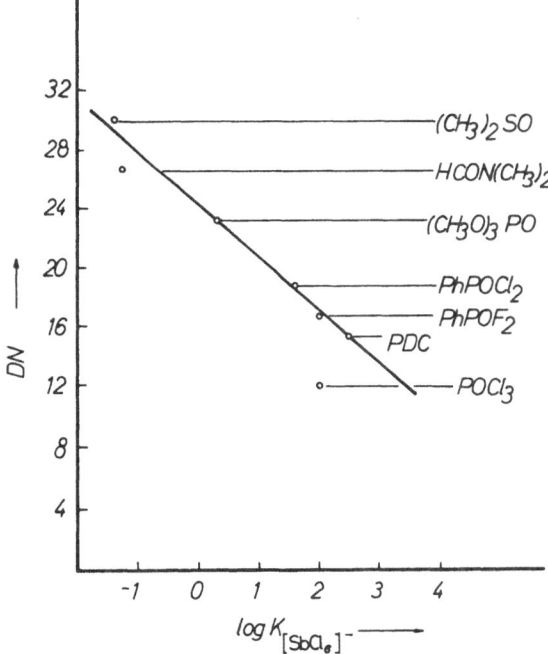

Fig. 10. Stability constants of $[SbCl_6]^-$ ions in solution of a donor solvent of given donicity $DN$.

$$D.SbCl_5 + Cl^- \rightleftharpoons [SbCl_6]^- + D$$

in the same solvent.

The stability constant of hexachloroantimonate

$$K_{[SbCl_6]^-} = \frac{a_{[SbCl_6]^-} \cdot a_D}{a_{D.SbCl_5} \cdot a_{Cl^-}}$$

in the respective solvent D is inversely proportional to the stability constant $K_{D.SbCl_5}$ and thus increases with a decrease in donicity of the solvent (Fig.10)[7]. Only in media of low donicity this relation is not obeyed, since $SbCl_5$ is partly associated in such solutions.

Preliminary results (obtained by use of the temperature jump technique) indicate that the reactions are fast (at $c \approx 10^{-2}$ relaxation time faster than 50 $\mu$sec.) [56] and are of second order. The rate coefficients $k_1$ are again a function of the donicity, whereas the rate coefficients of the reversed reactions $k_2$ show no relation to the donicity [56].

## 3.2. Complexes of Cobalt (II)

The following type of reaction may serve as an example for the formation of complex compounds of transition metal elements:

$$[CoD_6]^{2+} + 4 X^- \rightleftharpoons [CoX_4]^{2-} + 6 D; X^- = I^-, Br^-, Cl^-, NCS^- \text{ or } N_3^-$$

Such equilibria are known to consist of a number of consecutive complex equilibria. The formation constant $\beta$ represents the free enthalpy of complex formation in the gas phase. This quantity can not be determined by experiment.

Table 9 shows the molar ratios $\nu = c_{X^-}/c_{Co^{2+}}$, which are required in different solvents for the complete conversion of all cobalt (II) containing species into $[CoX_4]^{2-}$ units under analogous conditions.

The donor properties of the ligandes $I^-$, $Br^-$, $Cl^-$, $NCS^-$ and $N_3^-$ are much higher than that of nitromethane and the conversion into $[CoX_4]^{2-}$ is complete at the stoichiometric amounts of $X^-$ in this solvent. The stabilities of $[CoX_4]^{2-}$ – complexes for $X^- = Br^-$, $Cl^-$, $NCS^-$ and $N_3^-$ in acetonitrile, propanediol-1,2-carbonate, ethylene sulfite and acetone are in consonance with the donor properties of the solvents and the anionic ligands.

Iodide and bromide ions are considerably weaker donors than DMSO or HMPA: the formation of tetrahalocobaltate does not occur; the excess required

Table 9. *Excess of $X^-$ required for the complete formation of $[CoX_4]^{2-}$ in different donor solvents (1 represents the stoichiometric amount $X^-: Co^{2+} = 4$)*

| Donor DN | $X^-$ $I^-$ | $Br^-$ | $Cl^-$ | $NCS^-$ | $N_3^-$ | Ref. |
|---|---|---|---|---|---|---|
| NM (2.7) | 1.2 | 1 | 1 | 1 | 1 | 57, 58) |
| AN (14.1) | [3.2] | (10) | 4 | 2 | 2 | 43, 59−66) |
| PDC (15.1) | 2. | 1.2 | 1.2 | 1 | 1 | 43, 64, 65) |
| ES (15.2) | (22) | (11) | 4 | 2 | 2 | 67, 68) |
| Acetone (17.0) | 75 | 25 | 2 | 1.5 | 1 | 58, 69−73) |
| Water (18.0) | [750] | [750] | [750] | [750] | [750] | 72, 74−76) |
| DMA (27.8) | [82]$^a$ | [10]$^a$ | 4 | 3 | 1 | 43, 58, 65) |
| DMSO (29.8) | [10]$^c$ | | 50 | 50 | 5 | 43, 58, 65, 66, 77) |
| HMPA (38.8) | [48]$^b$ | [120]$^a$ | 115 | 40 | 7 | 78, 79) |

( ) = $[CoX_4]^{2-}$ not completely formed
[ ] = $[CoX_4]^{2-}$ not detectable
a) Equilibrium $CoX_2 - [CoX_3]^-$
b) Equilibrium $[CoX]^+ - CoX_2$
c) Equilibrium $Co^{2+} - [CoI]^+$

to obtain tetrathiocyanato cobaltate in DMA and DMSO is also in agreement with the donor properties. The high donor property of the azide ion is evidenced by the small amounts of azide ions necessary to form tetraazidocobaltate even in solvents of high donicity.

On the other hand, Table 9 contains several data which cannot be interpreted by the relative donicities of donor solvent and competitive ligand: the donor strength of the iodide ion is comparable to that of acetonitrile and the formation of tetraiodocobaltate in solvents of high donicity, such as DMA, DMSO and HMPA may not be anticipated. The formation of this species is however, complete with excess iodide ions and is due to the small values of the free enthalpies of the gas phase reactions:

$$[VO(acac)_2D] + I^- \rightleftharpoons [VO(acac)_2I]^- + D$$

In a donor solvent the iodide ions is much more strongly solvated than the neutral donor and hence the donor properties of the iodide ion are lowered in solution. This event has been described as the *thermodynamic solvatation effect*. It becomes increasingly important with an increase of the ratio of the free enthalpy of solvation to the free enthalpy of the ligand exchange reaction.

Since VO (acac)$_2$ is a weak acceptor the free enthalpy of the ligand exchange reaction is low and hence the thermodynamic solvation effect is considerable.

The hypothetical [CoCl$_3$]$^-$—ion is a stronger acceptor than VO(acac)$_2$ and hence the thermodynamic solvation effect with a given anion and in a given solvent is smaller. The decrease of the donor strength of an anion due to solvation is less apparent in the last step required for the formation of tetrahalocobaltate.

Indeed [CoCl$_4$]$^{2-}$ is formed in AN with fourfold excess of chloride ions, while a hundredfold excess of chloride ions is not sufficient for the complete formation of [VO(acac)$_2$Cl]$^-$ in the same solvent (Table 9). A similar situation is found by comparing [CoI$_3$]$^-$ and [Co(NCS)$_3$]$^-$ with VO(acac)$_2$.

For a given acceptor the relative donor strength of X$^-$ and the neutral donor are dependent on the solvating properties of the solvent toward anions. This has been termed *"specific solvation effect"* and it explains the following observations: According to Table 9 [CoX$_4$]$^{2-}$-complexes are formed in DMA more readily than in DMSO due to its higher solvating power towards anions. On the other hand, the tendencies to give [CoCl$_4$]$^{2-}$, [Co(NCS)$_4$]$^{2-}$ and [Co(N$_3$)$_4$]$^{2-}$ are similar in DMSO and in HMPA. The latter has a higher donicity, but is considerably weaker in solvating anions.

Despite the different donicities of ES and DMA the formation of [CoX$_4$]$^{2-}$ species is accomplished under similar conditions. This fact is again due to the differences in the solvating power towards anions. The higher donicity of DMA is outweighted by the higher solvating properties of ES; on the other hand, ES and PDC have similar donicities, but PDC is the weaker solvating medium.

The differences between acetone and water are drastic: the donicities are similar, but the solvating properties are vastly different: [CoX$_4$]$^-$ complexes are easily formed in acetone, but water prevents the formation to be completed even at considerable excess of the anion ligands. The azide ion is a very strong donor, but it is strongly solvated by water. For that reason only monoazidocobaltate and diazidocobaltate are formed in water [72]. Likewise, only small amounts of [Co(NCS)$_4$]$^{2-}$ are present in aqueous 8-molar KNCS solutions and [CoCl$_4$]$^{2-}$ and [CoBr$_4$]$^{2-}$ are formed only in concentrated solutions of the respective hydrogen halides.

## 3.3. Complexes of Nickel (II)

The acceptor properties of Ni$^{2+}$ are weaker than those of Co$^{2+}$ and under analogous conditions the extent of complex formation is smaller. When sodium azide is added to a $10^{-3}$ molar nickel (II) perchlorate solution in excess, the only azide-complex is monoazidonickel (II). The monochloro complex is present only in strong hydrochloric acid [81]. Due to the low solvating properties of DMA even tetrachloronickelate (II) is found in such solutions.

### 3.4. Steric Considerations

$Co^{2+}$ is known to give hexasolvated species with most donor molecules. With HMPA, however, the tetrasolvate $[Co(HMPA)_4]^{2+}$ is formed and the heat of solvation is smaller than would be expected by mere consideration of the donicity of HMPA. A plot of the enthalpies for the reaction

$$[Co(AN)_6]^{2+} + n\,D \rightleftharpoons [CoD_n]^{2+} + 6\,AN$$

vs. the donicities of D lie on a straight line for D = AN, DMF and DMSO, but this is not true for HMPA [83]. The $^1$H-NMR-spectra at different temperatures lead to the conclusion that a room temperature coordination center and ligand shell are not at rest. It seems that the space-requiring HMPA molecules form a rigid shell around the coordination center, which prevents the donor molecules to approach the former as close as possible. The coordination center appears to be "rattling" within the sphere of ligand molecules [83] and the Co-HMPA bonds are weaker than expected. Therefore two coordinated HMPA molecules are easily replaced by chloride ions to give $Co(HMPA)_2Cl_2$. Even iodide may replace at least one HMPA molecule [78].

$$[Co(HMPA)_4]^{2+} + I^- \rightleftharpoons [Co(HMPA)_3I]^+ + HMPA$$
$$[Co(HMPA)_4]^{2+} + 2\,Br^- \rightleftharpoons [Co(HMPA)_2Br_2] + 2\,HMPA$$

In the mixed complexes the "rattling-effect" is no longer observed and the bond energies of Co-HMPA are in accordance with the expectations.

### 3.5. Formation of Chelates

Certain metal ions react with trimethylphosphate to yield stronger complex compounds than would be expected from a mere consideration of its donicity. It has been found that this event is due to chelate formation [84, 85]:

$$87$$

### 3.6. Complexes of Soft Metal Ions

It has been mentioned that the donicity rule does not hold for typical soft-soft interactions. In particular, DMSO behaves differently towards soft metal ions and coordination is achieved through the "soft end" of the DMSO molecule. Whereas iodides of hard metal ions are completely ionized in DMSO, the iodides of soft metal ions tend to remain non-ionized in DMSO-solutions. $HgI_2$ is scarcely ionized in DMSO and shows a high tendency to react with iodide ions [54,86,87].

# 4. Ionization with Autocomplex Formation

### 4.1. General

The ionization of trimethyltin iodide by neutral donors is an example of a heterolytic fission of a covalent bond. The ionization process is more complicated if the substrate contains more than one ionizable bond, in particular, if the anions formed are capable of competing successfully with the donor molecules for coordination at the substrate. If they are successful both complex cations and complex anions are formed and this process is known as "autocomplex formation" or "ligand disproportionation"

Ionization with formation of a complex cation:    $D + MX \rightleftharpoons DM^+ + X^-$

Formation of a complex anion:    $X^- + MX \rightleftharpoons MX_2^-$

Autocomplex formation:    $D + 2\,MX \rightleftharpoons DM^+ + MX_2^-$

Autocomplex formation is favored when the substrate is difficult to ionize. In a strong donor solvent the substrate tends toward autocomplex formation if it is not easily ionized. On the other hand, a substrate which is easily ionized will not tend to give autocomplex ions in a strong donor solvent.

Frequently, adduct formation, ionization, and autocomplex formation, may occur simultaneously and to an extent which is governed by the stabilities of the resulting species. This process is also influenced by the molar ratios of donor and substrate, a factor which has frequently been ignored.

NMR measurements, spectrophotometric, kinetic, potentiometric, polarographic, and conductometric investigations, are helpful in elucidating the various types of coordination in solution. Conductometric titration in a coordinating inert medium of reasonable dielectric constant has proved to be very useful for obtaining indications about the superposition of autocomplex formation, adduct formation and ionization.

## 4.2. Tin (IV) Iodide

Autocomplex formation of tin (IV) iodide is indicated by the conductometric and spectrophotometric measurements [88]. Tin (IV) iodide gives a yellow non-conducting solution in nitrobenzene. The solution turns red upon addition of a donor: with strong donors, such as TBP, DMF, DMSO, or HMPA, the color change occurs on addition of the first drop; with weak donors considerable amounts may be necessary. The spectra show the presence of hexaiodostannate. At the same time the solution becomes conducting. The comparison of the conductivities at a molar ratio $D \cdot SnI_4$ $\nu = 3$ shows a relationship to the donicities of the neutral donors in the order

$$AN < THF < TBP < DMF < DMSO < HMPA$$

(Fig. 11). In general, the conductivities increase with increasing amounts of donor, but inflections are observed for D = DMF at $\nu \approx 1.9$, for D = DMSO at $\nu \approx 1.5$.

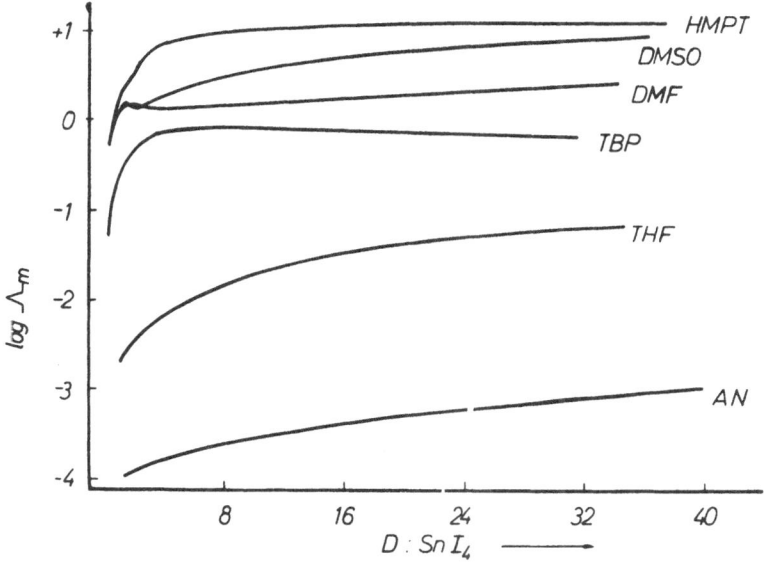

Fig. 11. Conductometric titrations of $SnI_4$ in nitrobenzene with different donors

The adduct $SnI_4D_2$ is partialy ionized by a small amount of a weak donor, but it is more strongly ionized by the same amount of a strong donor:

$$SnI_4D_2 + D \rightleftharpoons [SnI_3D_3]^+ + I^- \tag{1}$$

89

At the same time the iodide ions formed by ionization are capable of forming hexaiodostannate:

$$SnI_4D_2 + 2I^- \rightleftharpoons [SnI_6]^{2-} + 2D \qquad (2)$$

and hence autocomplex formation occurs:

$$3 Sn_4D_2 \rightleftharpoons 2 [SnI_3D_3]^+ + [SnI_6]^{2-} \qquad (3)$$

The extent of autocomplex formation depends on both, the nature and the quantity of the donor added. For a given molar ratio the autocomplex formation is decreased by an increase in donicity, since the iodide ions are less successful in competing with an excess of donor molecules. At the same time the donor molecules replace iodide coordination in hexaiodostannate:

$$[SnI_6]^{2-} + 2D \rightleftharpoons SnI_4D_2 + 2I^- \qquad (4)$$

and the iodide ions may react with the complex cations:

$$[SnI_3D_3]^+ + I^- \rightleftharpoons SnI_4D_2 + D$$

In this manner the number of ions is decreased and the conductivity curve shows an inflection: In DMF and DMSO even a flat maximum is found, which is followed by a flat minimum. Upon further addition of the donor the red color disappears and the conductivity increases steadily, since ionization according to Eq. (1) is now predominating.

The inflection is hardly pronounced for D = HMPA since the donor is strong and formation of $[SnI_6]^{2-}$ takes place only to a small extent. The flat maximum in the conductivity curve for D = TBP is due to a physical effect: a decrease in $\epsilon$ and an increase in viscosity. There is no inflection for THF and the solution remains red even at high donor contents of the solution, since autocomplex species are the most stable entities in this system.

No experimental evidence is available to substantiate the composition of the cationic species. In dilute $SnI_4$ solutions and in the presence of large amounts of a donor of high donicity $[D_4SnI_2]^{2+}$, $[D_5SnI]^{3+}$, and even $[D_6Sn]^{4+}$, might be existent [54].

## 4.3. Iron (III) Chloride

The ionization of iron (III) chloride is less readily achieved than that of tin (IV) iodide. Iron (III) chloride gives a nonconducting solution in dichloroethane and addition of a donor does not cause the formation of a conducting solution below the molar ratio D:FeCl$_3$ = 1:1. Upon further addition of the donor the solution becomes conducting and turns yellow due to the formation of tetrachloroferrate [89].

$$D + FeCl_3 \rightleftharpoons D.FeCl_3 \text{ (non-conducting)}$$
$$2 D + 2 D.FeCl_3 \rightleftharpoons [D_4FeCl_2]^+ + [FeCl_4]^-$$

This mode of autocomplex formation is observed to a slight extent in a solution of iron (III) chloride in nitrobenzene. The low conductivity of this solution is increased according to the amount and donicity of the donor added (Fig. 12).

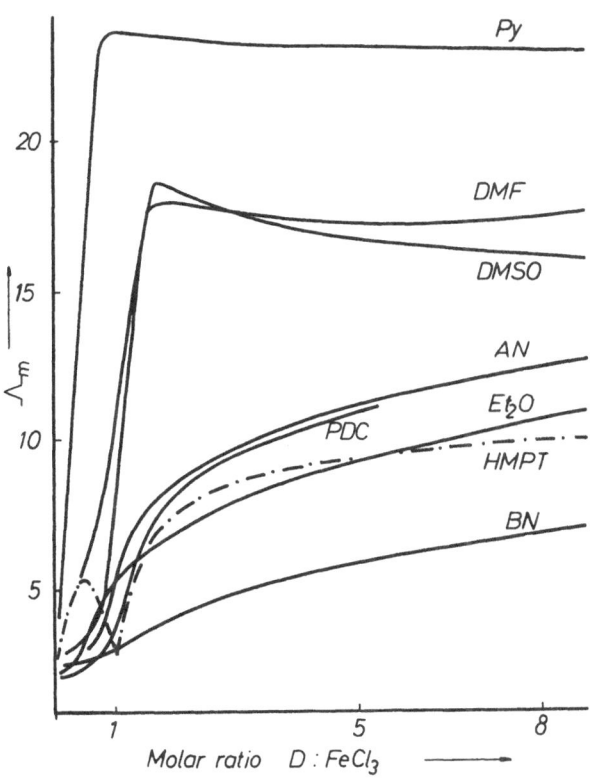

Fig. 12. Conductometric titrations of FeCl$_3$ in nitrobenzene with different donor molecules D

HMPA gives, however, poorly conducting solutions [89]. In the course of the conductometric titration of $FeCl_3$ with HMPA in nitrobenzene a conductivity maximum is observed at a molar ratio HMPA : $FeCl_3$ = 1 : 2 and $[FeCl_4]^-$ ions are present at this composition of the solution. It is likely that the complex cation which is simultaneously produced by autocomplex formation may contain coordinated nitrobenzene molecules:

$$3 \text{ NB} + \text{HMPA} + 2 \text{ FeCl}_3 \rightleftharpoons [(NB)_3 HMPAFeCl_2]^+ + [FeCl_4]^-$$

Upon further addition of HMPA the conductivity of the solution drops considerably and passes through a minimum at a molar ratio 1:1. Apparently, the molecular adduct is the most stable species under these conditions:

$$[FeCl_4]^- + \text{HMPA} \rightleftharpoons \text{HMPAFeCl}_3 + Cl^-$$
$$[HMPAFeCl_2]^+ + Cl^- \rightleftharpoons \text{HMPA.FeCl}_3$$

Further addition of HMPA leads to an increase in the conductivity and the $[FeCl_4]^-$ ions are no longer found to be present. The ionization equilibrium

$$4 \text{ DMSO} + 2 \text{ FeCl}_3 \rightleftharpoons [(DMSO)_4 FeCl_2]^+ + [FeCl_4]^-$$

and hence the conductivity of the HMPA-containing solution is considerably smaller than that of the DMSO-solution, where autocomplex formation occurs.

There is a slight decrease in the amount of ions present once a large excess of DMSO is added as is evidenced by the slight decrease in conductivity (which is also observed with D = pyridine). It is impossible to conclude from these results, whether or not in the presence of large amounts of donor the autocomplex equilibrium is superimposed by a simple ionization equilibrium. Since DMSO has a higher tendency to solvate anions than HMPA it appears likely that, at least to some extent, simple ionization is involved [89].

On the other hand, the conductivity is rising in the presence of an excess of a weak donor, since the autocomplex equilibrium is far on the side of the adduct and excess of donor will produce more ions. Simple ionization with formation of chloride ions seems less likely in these cases.

## 4.4. Boron (III)-Chloride

The solution of boron trichloride in nitrobenzene does not conduct the electric current. When a donor is added, the conductivity rises sharply up to a molar ratio $D:BCl_3 = 1:1$ (Fig. 13), which is considered due to autocomplex formation:

$$2 \ D.BCl_3 \ \rightleftharpoons \ [D_2 BCl_2]^+ + [BCl_4]^-$$

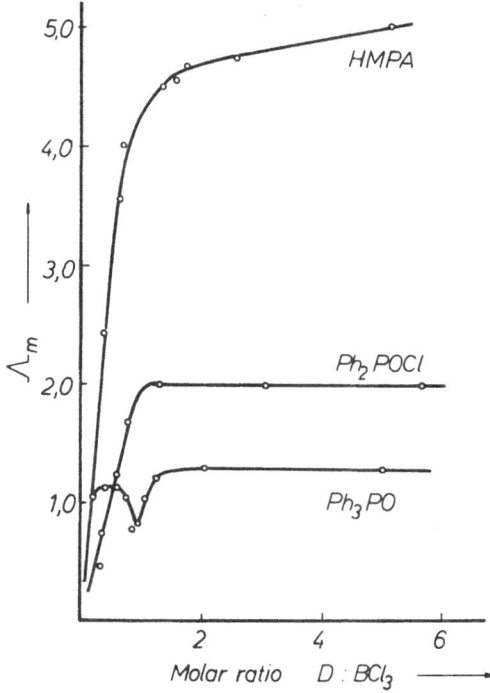

Fig. 13. Conductometric titrations of $BCl_3$ in nitrobenzene with different P=O-donors

For the system $Ph_2POCl$-$BCl_3$ NMR evidence is available to support this assumption. The [31]P NMR spectra indicate the presence of free $Ph_2POCl$ ($\delta_P$ −42.7 ppm), of the adduct $Ph_2POCl.BCl_3$ ($\delta_P$ −61.2 ppm) and of a further species ($\delta_P$ −64.2 ppm), which may be considered as the complex cation $[(Ph_2POCl)_2 BCl_2]^+$: at higher $BCl_3$ concentrations the intensity of this signal increases and those of the other signals are decrease [90].

The molar conductivities at $\nu = 1:1$ are increased according to the donicities of the donors, namely

$$PhPOF_2 \ < \ PhPOFCl \ < \ PhPOCl_2 \ < \ Ph_2POCl \ < \ HMPA$$

An exception is provided by triphenylphosphine oxide.

The conductivity curve is analogous to that of the system HMPA-FeCl$_3$ and the conductivities are lower than those in the presence of Ph$_2$POCl, which is a weaker donor than Ph$_3$PO (Fig. 13). This observation is interpreted by analogy to the HMPA-FeCl$_3$ system. The autocomplex formation

$$Ph_3PO + 2\ BCl_3 \rightleftharpoons [(Ph_3PO)BCl_2]^+ + [BCl_4]^-$$

is considerably suppressed at a molar ratio Ph$_3$PO:BCl$_3$ = 1 : 1 due to the higher stability of the molecular adduct, which at higher concentrations of BCl$_3$ is moderately ionized:

$$Ph_3PO + (Ph_3PO)BCl_3 \rightleftharpoons [(Ph_3PO)_2BCl_2]^+ + Cl^-$$

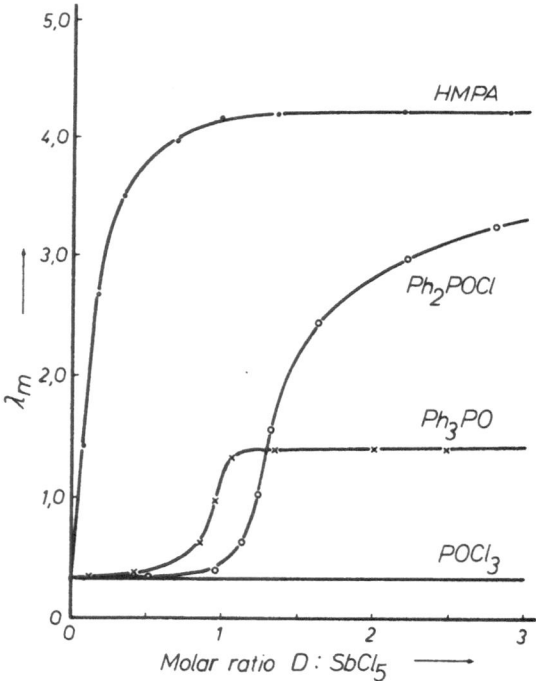

Fig. 14. Conductometric titrations of SbCl$_5$ in nitrobenzene with different P=O-donors

It is likely that with HMPA the simple ionization is dominating even at low molar ratios.

NMR investigations of the $BCl_3$ systems with TMP and TBP indicate the formation of chelates and of alkyl chloride.

## 4.5. Antimony (V)-Chloride

Antimony pentachloride is known to have a small tendency towards ionization and this is reflected in the conductivities, which are lower than those of the corresponding $BCl_3$-systems. Donor molecules of donicities below 15 give non-conducting solutions even at high donor contents.

With the exception of HMPA strong neutral donors give non-conducting solutions; up to a molar ratio $D:SbCl_5 = 1:1$ further addition of the donor increases the conductivity apparently due to autocomplex formation:

$$2 \, D.SbCl_5 \; \rightleftharpoons \; [D_2SbCl_4]^+ + [SbCl_6]^-$$

The final conductivity of the $Ph_3PO$ system is lower than that of the $Ph_2POCl$ system, although the former is known to have a higher donicity than the latter. Cryoscopic measurements in the $Ph_2POCl$ system reveal that the number of particles is increased by the ionization process and thus autocomplex formation can be excluded. The following mode of ion formation has been suggested:

$$Ph_2POCl + SbCl_5 \; \rightleftharpoons \; [Ph_2PO]^+ + [SbCl_6]^-$$

In the $HMPA-SbCl_5$ system the conductivity curve is similar to that in the $HMPA-SnI_4$ system. Since $SbCl_5$ is much more difficult to ionize, autocomplex formation is more likely than simple ionization:

$$2 \, HMPA.SbCl_5 \; \rightleftharpoons \; [HMPA)_2SbCl_4]^+ + [SbCl_6]^-$$

## 4.6. Trihalides of Arsenic and Antimony

The trihalides of arsenic and antimony are known to act as acceptor molecules and to form adducts with various donor molecules. The complexes formed with 2,2'-bipyridyl are appreciably ionized in nitrobenzene [91] and simple ionization has been proposed [91, 92]:

$$MX_3 + bipy \; \rightleftharpoons \; [(bipy)MX_2]^+ + X^-$$

95

The latter increases from the chloride to the iodide and from antimony to arsenic. As expected the $\Delta H$-values increase with an increase in donicity of the neutral donor [9] and also from chloride to iodide, but they are lower for arsenic than for antimony [9].

This observation suggests that the ionization process is endothermic. Conductometric titrations of the trihalides with HMPA in 1,2-dichloroethane suggest that at low $D:MX_3$ ratios some autocomplex formation may occur. At a molar ratio of 1:1 inflections are found indicating that the mode of the ionic equilibrium is essentially changed, apparently to that of simple ionization (Fig. 15)

$$2 D + MX_3 \rightleftharpoons [D_2MX_2]^+ + X^-$$

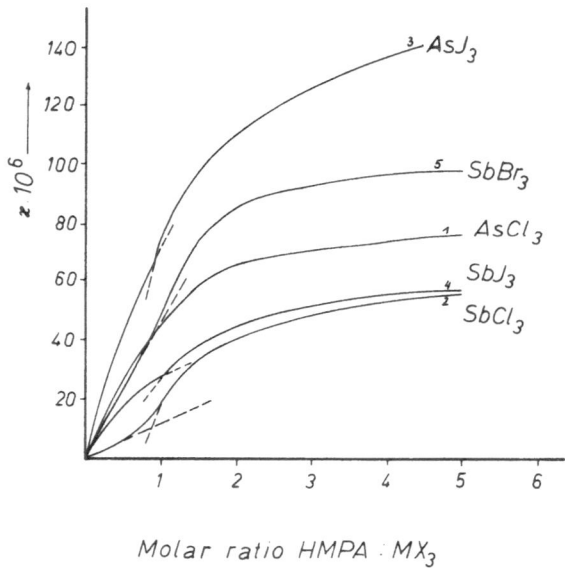

Fig. 15. Conductometric titrations of various group-V-trihalides with HMPA in nitrobenzene

This interpretation is in agreement with the results of a recent Raman spectroscopic study [93] supporting the absence of $[MX_4]^-$ — ions and also with the results of other spectrophotometric investigations [9]. In the system $AsI_3$-HMPA the final spectrum is obtained at a molar ratio $D:AsI_3 = 2:1$, whereas with the weaker donor DMA a considerable excess of donor is required in order to attain the final spectrum.

## 4.7. Carbonyl Compounds

Metal carbonyls are subject to autocomplex formation in the presence of strong donor molecules [94-98]. Besides the cation which is coordinated by donor molecules, polynuclear anions are formed; the latter can be degradated at higher temperatures. It may be noted that in this process of autocomplex formation changes in the oxidation numbers and thus redox reactions are involved:

$$4\ Fe(CO)_5 + 6\ py \rightleftharpoons [Fe(py)_6]^{2+}[Fe_3(CO)_{11}]^{2-} + 9\ CO$$

The conductivity of pyridine is considerably increased by addition of iron carbonyl and corresponds to that of potassium picrate in the same solvent [97]. Likewise, DMSO, alcohols, or ammonia [95, 99], as well as soft ligands such as triphenylphosphine, may serve as donor molecules [100]:

$$3\ Co_4(CO)_{12} + 24\ D \rightleftharpoons 4\ [CoD_6][Co(CO)_4]_2 + 4\ CO$$
$$Co_6(CO)_{16} + 12\ D \rightleftharpoons 2\ [CoD_6][Co(CO)_4]_2$$
$$Co_2(CO)_8 + PPh_3 \rightleftharpoons [Co(CO)_3(PPh_3)]^+[Co(CO)_4]^- + CO$$

# 5. Donicity and Rates of Solvent Substitution

## 5.1. The Equilibria $DMX_n + X^- \rightleftharpoons [MX_{n+1}]^- + D$

### 5.1.1. For a Given Acceptor in Different Solvents

The strength of a solvent bond influences the rate of solvent substitution in a given compound. Kinetic measurements by means of the T-jump relaxation technique have illustrated that for the reactions of the solutions of $SbCl_5$ with triphenylchloromethane in different solvents a relationship exists between the rate constant and the donicity of the solvent used.

$$D.MCl_n + Ph_3CCl \underset{k_{21}}{\overset{k_{12}}{\rightleftharpoons}} [Ph_3C]^+[MCl_{n+1}]^- + D$$

The logarithm of the rate coefficient is a linear function of the donicity of dichloromethane, nitromethane, benzonitrile, and acetonitrile (Fig. 16). Deviations in the $DN - \log k_{12}$ plot are found for solvents with viscosities $> 2$ cP. The deviation is noticeable for solvents of viscosities between 2 and 3 cP (TMP 2.32 cP, PDC 2.83 cP) and is more pronounced for a solvent of considerably higher viscosity such as phenylphosphonic dichloride (4.10 cP) (Fig. 16).

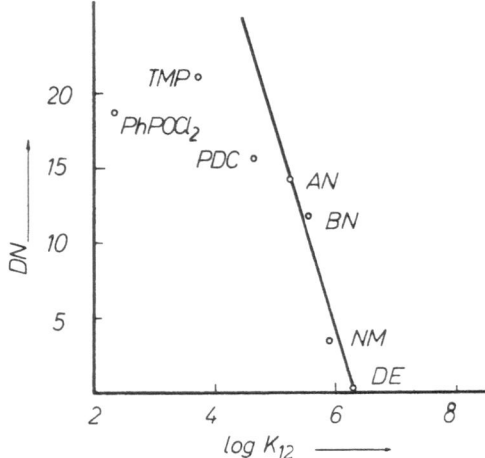

Fig. 16. Donicity and rate constant $k_{12}$ for the reaction $Ph_3CCl + D.SbCl_5 \rightleftharpoons [Ph_3C]^+$ $[SbCl_6]^- + D$ in donor solvents D of different donicity

The influence of the donicity on the rate of the reverse reaction $k_{21}$ is considerably smaller, probably since the ratedetermining step is the dissociation of a chloride ion ($SN_1$ mechanism).

Table 10. *Rate coefficients $k_{12}$ and $k_{21}$ for the reaction*
$D.SbCl_5 + Ph_3CCl \rightleftharpoons [Ph_3C]^+ [SbCl_6]^- + D$ in DE and AN

| Solvent | DN | $k_{12}$ | $k_{21}$ |
|---------|-----|----------|----------|
| Dichlorethane | 0 | $1.75 \cdot 10^6$ | 5 |
| Acetonitrile | 14.1 | $1.77 \cdot 10^5$ | 17.3 |

While $k_{12}$ in AN is tenfold higher than in DE, $k_{21}$ in AN is only three times its value in DE. These findings are in accordance with the relationship between the formation constant of hexachloroantimonate and the donicity of the utilized solvent, as has been stated in Sect. 3.1. The values for the equilibrium constants $K_{[SbCl_6]^-}$ obtained from the kinetic measurements are in agreement [102] with those found from equilibrium studies[4, 7] in the respective solvents (Fig. 17). In solvents of very low donicity the $K_{[SbCl_6]^-}$ values are lower than expected on the basis of the DN-$K_{[SbCl_6]^-}$ plot. This may be attributed to the presence of polymeric $SbCl_5$ units, and thus to the involvement of a second equilibrium.

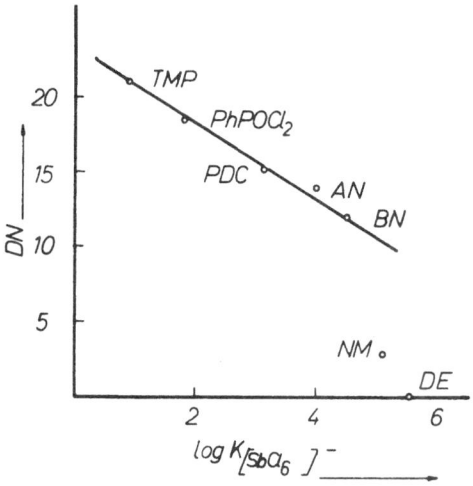

Fig. 17. Stability constants $K[SbCl_6]^-$ in donor solvents of different donicity

## 5.1.2. Different Acceptors in a Given Solvent

Similar behavior is shown by $GaCl_3$, $InCl_3$ and $FeCl_3$ as acceptors which have acceptor properties similar to those of $SbCl_5$ [103] as indicated in Table 11.

Table 11. *Rate coefficients* $k_{12}$ *and* $k_{21}$ *for the reactions*
$D.MCl_n + Ph_3CCl \rightleftharpoons [Ph_3C]^+[MCl_{n+1}]^- + D$ *in AN*

| Acceptor Chloride | $k_{12}$ | $k_{21}$ |
|---|---|---|
| $SbCl_5$ | $1.77.10^5$ | 17.5 |
| $GaCl_3$ | $2.75.10^5$ | 20 |
| $InCl_3$ | $2.72.10^5$ | 18.5 |
| $FeCl_3$ | $1.22.10^5$ | 24.5 |

## 5.2. Substitution in Hydrated Metal Ions in Water

Kinetic measurements for reactions of the type

$$[MD_6]^{n+} + L \underset{k_{21}}{\overset{k_{12}}{\rightleftharpoons}} [MD_5L]^{n+} + D$$

have been made in aqueous media only [104-106]. A relationship exists between the strength of the solvate bond and log $k_{12}$ a nearly straight line is obtained for the alkali metal ions and alkaline earth metal ions, as well as for certain transition elements of equal charge (Fig. 18 and 19).

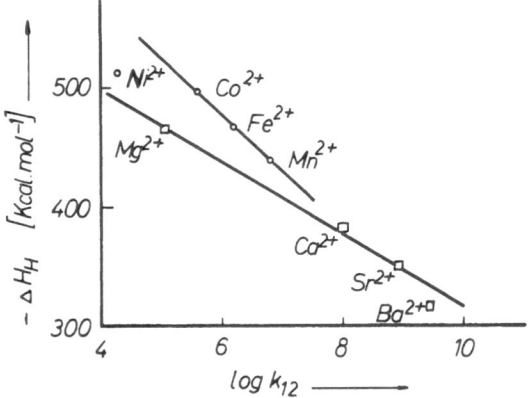

Fig. 18. Rate constants $k_{12}$ and enthalpy of hydration for the solvent substitution reactions in water for various metal ions

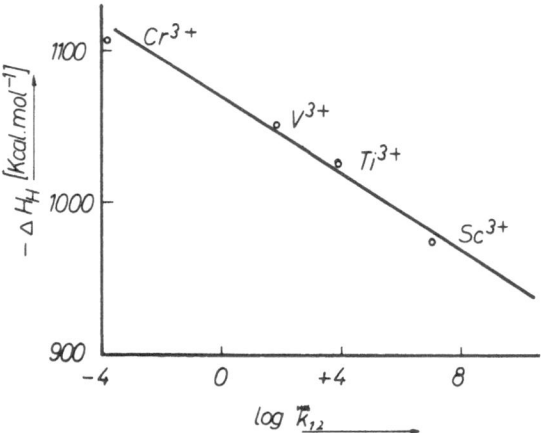

Fig. 19. Rate constants $k_{12}$ and enthalpy of hydration for the solvent substitution reactions in water for various metal ions

The plot supplies similar informations as the plot of the reciprocal value of ionic radius of the metal $vs\ k_{12}$ described by Eigen and coworkers [107-109]. Deviations are considered to be due to ligand field stabilization, the Jahn-Teller effect, or to the effective ionic charge [110]. The rate coefficient of hydrated $Cu^{2+}$ is higher by three orders of magnitude as compared to that of most other elements: due to the Jahn-Teller effect the distances of the two water molecules in axial positions are much larger than those of the four water molecules in equatorial positions [111].

# 6. Donicity and Solvation of Metal Ions

Repeated reference has been made to solvation of metal ions and – since the donicity represents a $\Delta H$-value – it was surprising to note that in many instances the donicity of donor solvent molecules served as a useful guide for the interpretation of the behavior of metal solvates. The donicity does not take into account

a) entropic contributions to the $\Delta G$-values,

b) steric factors, which may be substantial for bulky solvent molecules and for small metal ions (Sect. 3.4), and

c) different bonding of solvent molecules within one coordination sphere, as is known to be due to the Jahn-Teller effect (Sect. 5.2).

Furthermore, the applicability of the donicity rule may be unexpected for the solvation of alkali metal ions, where a complete explanation of the observations may be provided by considering electrostatic interactions between ion and dipolar solvent molecules.

Table 12. *Enthalpies of hydration $\Delta H_H$ and $pK_S$-values for pairs of aquo-ions of equal charge and similar ionic radius*

| Aquoion | Electron configuration | Property | $r^\circ$ [Å] | $- \Delta H_H$ [kcal/mol] | $pK$ in water at $25^\circ$ |
|---|---|---|---|---|---|
| $K^+$ | $[Ar]$ | Hard | 1.33 | 73 | 14 |
| $Ag^+$ | $[Kr]4d^{10}$ | Soft | 1.26 | 112 | 10 |
| $Mg^{2+}$ | $[Ne]$ | Hard | 0.69 | 437 | 12.2 |
| $Cu^{2+}$ | $[Ar]3d^9$ | Soft | 0.69 | 499 | 7.3 |
| $Ca^{2+}$ | $[Ar]$ | Hard | 0.99 | 362 | 12.6 |
| $Cd^{2+}$ | $[Kr]4d^{10}$ | Soft | 0.97 | 428 | 9.0 |
| $Sr^{2+}$ | $[Kr]$ | Hard | 1.13 | 327 | 13.1 |
| $Hg^{2+}$ | $[Xe]5d^{10}$ | Soft | 1.10 | 443 | 3.6 |

It has been pointed out that various alkali metal salts in water are more asso-ciated than is expected on the basis of the electrostatic theory and also that metal ions of equal size and equal charge but of different electronic structure exhibit different enthalpies of hydration and different acidity constants in water [11].

The transition metal ions have higher hydration enthalpies and higher acidity constants than $s^2p^6$ ions of corresponding size and charge. Thus for the $d$-ions, covalent bonding is involved to a larger extent than for the ions of the alkali or alkaline earth metal ions. Stronger covalency means stronger polarization of the metal-oxygen bonds and weakening of the O—H bonds; on that basis deproto-nation is facilitated.

It has been stressed that it is useful to consider the enthalpy of solvation for the solvation of several metal ions. Erlich, Roach and Popov [112] investigated the $^{23}$Na NMR chemical shifts of NaClO$_4$ and NaBF$_4$ in different donor sol-vents and found the chemical shifts to be independent of the salt concentrations. Apparently, changes in the chemical shift of the cation are due only to changes in the shielding brought about by the formation of the inner solvation shell. Plotting the chemical shift $vs$ donicities (Fig. 20) gives approximately a straight line (the only discrepancy is water). Thus ionic solvation is appreciably influen-ced by the donor properties of the solvent molecules rather than by the dielec-tric constant. It may be concluded that in solvation of such hard ions (hydration is possibly an exception) covalent bonding is at least of some significance.

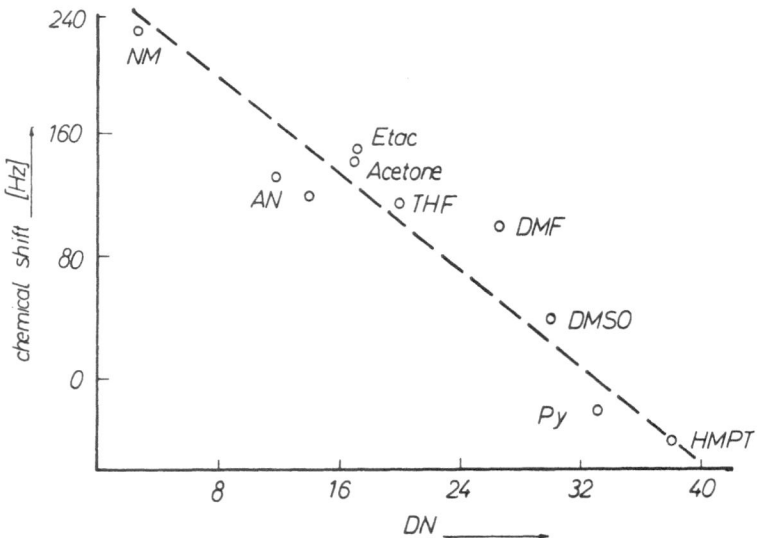

Fig. 20. $^{23}$Na NMR results: chemical shifts $vs$. donicity of the donor solvent D

# 7. Donicity and Redox Equilibria

## 7.1. General

Interesting results have been obtained from *polarographic studies* in various donor solvents. Measurements have been made of various metal perchlorates in solutions of donor solvents containing tetraalkylammonium perchlorate as supporting electrolyte against an aqueous saturated calomel electrode [113]. In order to eliminate differences in liquid-liquid junction potentials bisbiphenylchromium (I) has been used as a reference ion [114-118].

The relationship between the free enthalpy $\Delta G^\circ$ for the reaction

$$M_S \rightleftharpoons M_{SV}^{+n} + n\,e^-$$

and the standard electrode potential $E^\circ$ for the system $M_S/M_{SV}^{n+}$ is given by the equation

$$\Delta G^\circ = -n\,FE^\circ$$

In order to estimate the standard electrode potential of a metal ion a Born-Haber cycle consisting of the following three steps may be considered [119, 120] (Fig. 21):

        (a) sublimation of the metal,
        (b) ionization of the gaseous metal atom,
        (c) solvation of the cation.

For a given metal ion the free enthalpies for (a) and (b) remain constant and hence the value for $E^\circ$ of a given redox system in various solvents is determined by the corresponding free enthalpies of solvation of the cation.

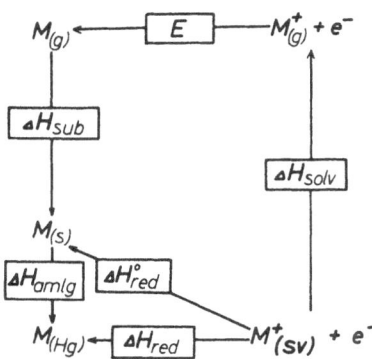

Fig. 21. Born-Haber cycle for the reduction of a solvated metal ion to the metal at a mercury electrode

$$\Delta G^\circ = \Delta H^\circ - T\Delta S^\circ$$

As a first approximation to entropic terms $\Delta S^\circ$ in the respective solvent may be considered as constant; $E^\circ$ for a given ion in different solvents is then determined by the enthalpy of solvation $\Delta H^\circ$.

For metals which are soluble in mercury, such as the alkali and alkaline earth metals, the polarographic half-wave potential is a function of the

      (a)standard electrode potential of the metal-metal ion complex,

      (b)solubility of the metal in mercury, and

      (c)free enthalpy of amalgamation.

Since (b) and (c) are independent of the nature of the solvent, the half-wave potential for a reversible reduction is a measure of the amount of interaction of the metal ion with the solvent molecules according to the reaction

$$M_{SV}^{n+} + n\,e^- \rightleftharpoons M\,(Hg) + \text{solvent}$$

provided the following conditions are fulfilled [120]:

      (a) absence of complex formation of the depolarizer,

      (b) absence of associated ions involving the depolarizer,

      (c) knowledge of activity coefficients, and

      (d) elimination of diffusion potentials.

For an irreversible reduction the half-wave potential is determined not only by the standard electrode potential but also by the polarographic overvoltage. For a simple electrode process the metal ion-solvent interaction is mainly responsible for the polarographic overvoltage and hence $E_{1/2}$ of such irreversible reductions may also be considered as a function of the solvation [119].

## 7.2. Donicity and Half-wave Potential for a Given Redox Couple

In the absence of data for the solvation enthalpies in nonaqueous media, an attempt has been made to plot the half-wave potentials for a given ion in different solvents (expressed in the bis(biphenyl)chromium(I) scale) *vs* the donicity of the solvent molecules. The Fig. 22 to 26 reveal a relationship between $E_{1/2}$ and *DN*, i.e., the half-wave potential becomes more negative with increasing donicity of the solvent.

In this plot a characteristic curve is found for each ion. The half-wave potentials of $Na^+$, $K^+$, $Rb^+$ and $Cs^+$ (Fig. 22) are similar in each of the solvents. The curve in the $E_{1/2}$-*DN* diagram reveals that in strong donor solvents $E_{1/2}$ remains nearly constant at increasing donicity [120, 121]. This observation suggests that these ions cannot utilize the strong donor properties of such solvents and that solvation is mainly due to electrostatic forces between ion and solvent dipoles.

This interpretation implies a high degree of order outside the inner coordination sphere for which dipole-dipole interactions between solvent molecules may be responsible. When the solvated metal ion is reduced at the dropping mercury electrode, it is completely desolvated to give an atom in the amalgam phase and the ordered solvation shell is broken down. This process accounts for $+\Delta S$ terms in the process of reduction. The lithium ion is found to be reduced at a more negative potential in the respective solvent [120, 121] (Fig. 22). There also exists a dependency of $E_{1/2}$ on the donicity in strong donor solvents suggesting more covalent bonding in $Li^+$ solvates, than in $Na^+$ solvate complexes.

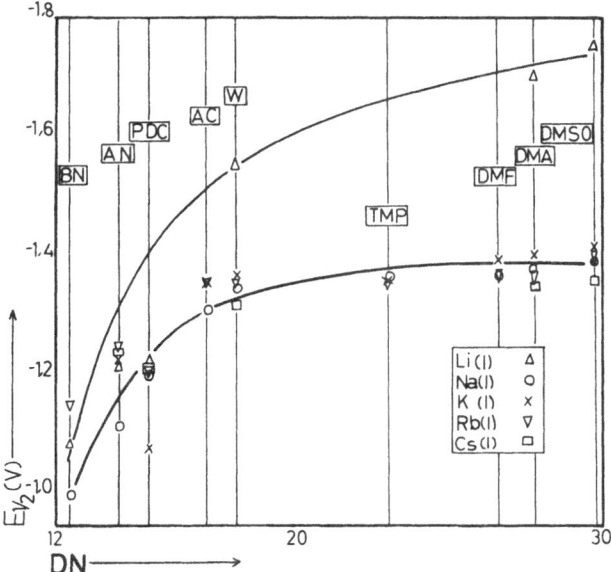

Fig. 22. $E_{1/2}$ and donicity of D for the alkali metal ions

The curves for $Ca^{2+}$, $Sr^{2+}$ and $Ba^{2+}$ are analogous to that for $Li^+$ and $E_{1/2}$ becomes more positive in this order (Fig. 23). It is interesting to note that $Sr^{2+}$ is reduced at a more negative potential than $K^+$ in a strong donor solvent, where-as $E_{1/2}$ for $K^+$ is more negative in solvents of donicities below 18. Thus in DMF, DMA or DMSO alkali metal ions are reduced by Sr

$$Sr + 2 K^+ \rightleftharpoons Sr^{2+} + 2 K$$

but $Ba^{2+}$ is more positive than $K^+$ in any of the investigated solvents. The ions $Yb^{2+}$, $Eu^{2+}$ and $Sm^{2+}$ give similar curves in the $E_{1/2}$-$DN$ plot [120] namely bet-

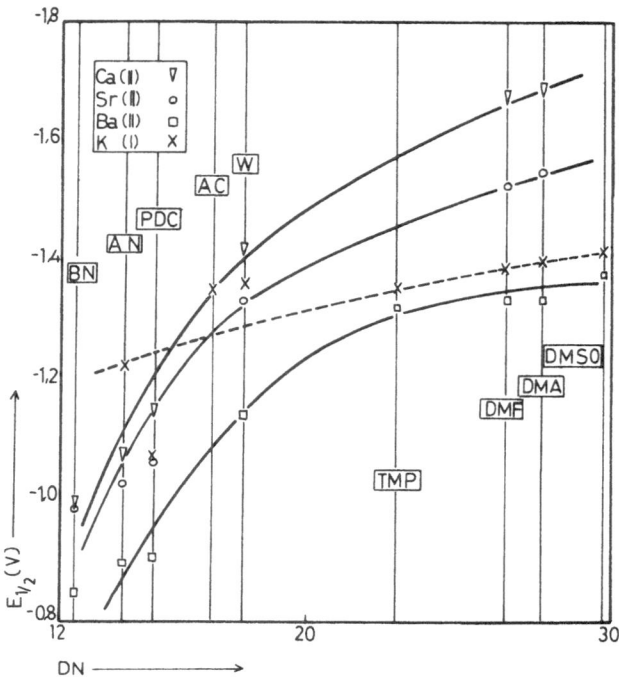

Fig. 23. $E_{1/2}$ and donicity of D for the alkaline earth metal ions

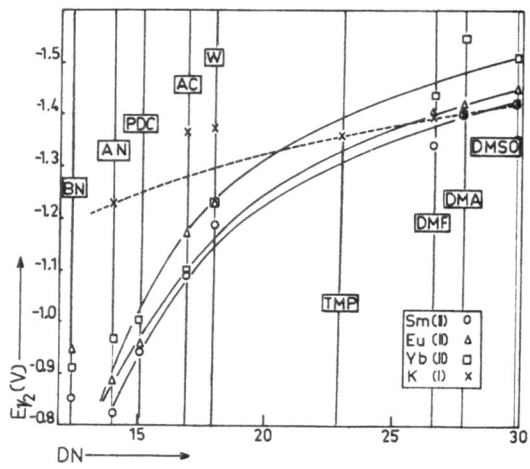

Fig. 24. $E_{1/2}$ and donicity of D for Sm(II), Eu(II) and Yb(II)

ween those of $Sr^{2+}$ and $Ba^{2+}$. Ytterbium appears to be capable of reducing the alkali metal ions (expect $Li^+$) in solutions of DMF, DMA and DMSO (Fig. 24).

$Zn^{2+}$ and $Cd^{2+}$ give a nearly straight line in the $E_{1/2}$-$DN$ plot (Fig. 25).

For $Ni^{2+}$, $Co^{2+}$ and $Tl^+$ increasing donicity of the solvent has a more pronounced effect on the half-wave potential (Fig. 26); the solvate bonds become increasingly covalent with strong donor molecules.

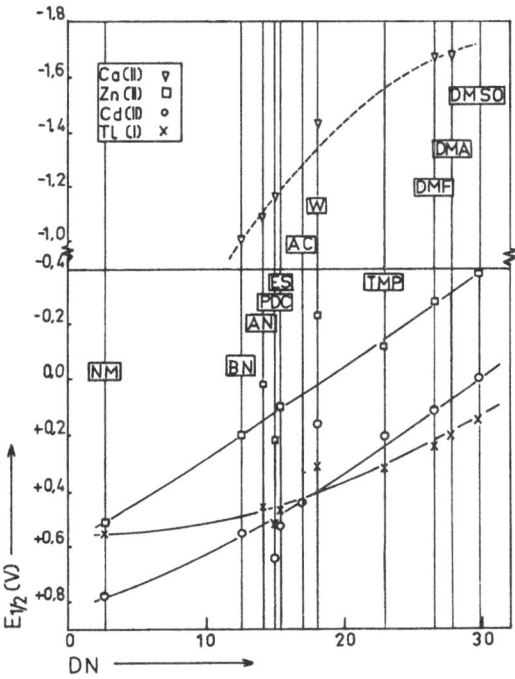

Fig. 25. $E_{1/2}$ and donicity of D for $Zn(II)$, $Cd(II)$ and $Tl(II)$

Several points are observed outside the curves, for several ions in trimethylphosphate and water.

Certain transition metal ions such as $Co^{2+}$, $Ti^{3+}$ are known to form chelates with trimethylphosphate, i.e., dimethoxyphosphato complexes [84, 85]. The chelate effect is responsible for the high stabilities of such complexes, which is expressed in the more negative values for the half-wave potentials. All ions producing such complexes are expected to undergo reduction in TMP at more negative potentials than would be expected from interpolation of the curves.

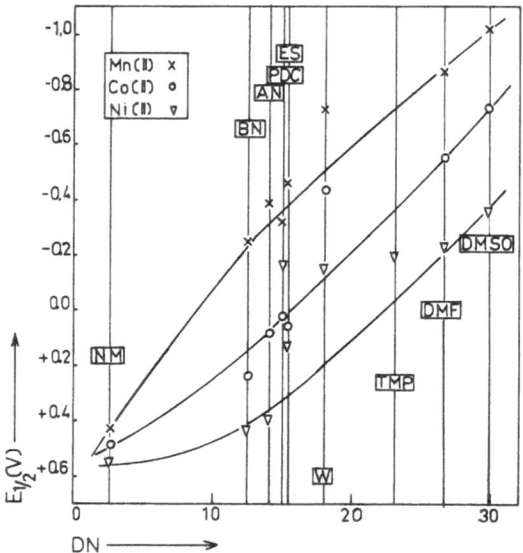

Fig. 26. $E_{1/2}$ and donicity of D for Mn(II), Co(II) and Ni(II)

The half-wave potentials of $K^+$, $Tl^+$ and $Ca^{2+}$ in water are slightly more negative and those for $Zn^{2+}$, $Cd^{2+}$, $Mn^{2+}$, $Ni^{2+}$ and $Co^{2+}$ considerably more negative than is expected according to the donicity rule. It has been shown in the previous sections that water is a rather unique solvent. The effect in question may be interpreted by the so-called "Katzin-effect" according to which water forms a "royal core of coordinated water molecules" which are hooked together by hydrogen bonds [70, 71, 122, 123].

Although no data are available in HMPA, it has been shown that due to steric effects metal ions are weaker coordinated than would be expected from its donicity [83]. This observation suggests that the half-wave potentials will be found at more positive potentials than expected from extrapolation of the curves.

In most other cases the relationship will allow the approximate prediction of the half-wave potentials of a given ion in a solvent of given donicity by interpolation. It may be expected that $E_{1/2}$ for a certain metal ion in tetramethylene sulfone ($DN$ = 14.8) will be similar to that in PDC ($DN$ = 15.1), benzylcyanide ($DN$ = 15.1) or ethylene sulfite ($DN$ = 15.3). Likewise, the half-wave potentials are expected to be similar in nitrobenzene ($DN$ = 4.4) and nitromethane ($DN$ = 2.7). In an analogous manner the half-wave potentials may be predicted in methyl acetate, diethylether, pyridine, and various other solvents.

The compilation of such data may further be used to tabulate electromotive series of redox potentials in any given donor solvent.

So far, the reduction of metal ions into the metallic state was discussed involving a complete removal of the coordinated solvent molecules in the reduction process. We shall now consider such redox-systems in which both the oxidized and the reduced species are solvated. The polarographic reduction of Eu(III) to Eu(II) in different solvents occurs at such half-wave potentials which are again related to the donicity of the solvent molecules [118]. In the $E_{1/2}$-DN plot a straight line is observed. Analogous results were obtained for the redox complexes Sm(III)-Sm(II) and Yb(III)-Yb(II) [118, 120] (Fig. 27).

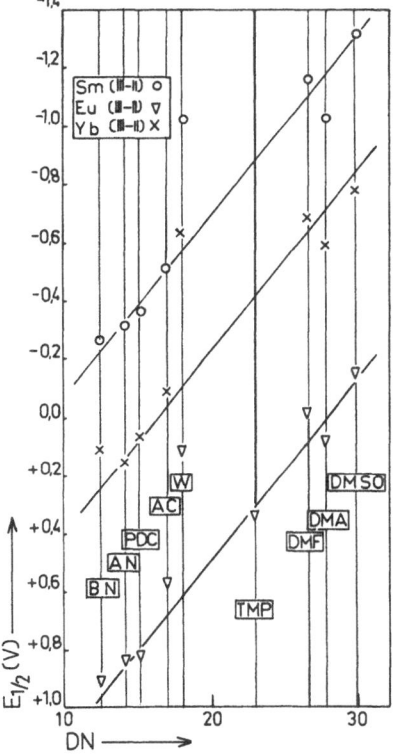

Fig. 27. $E_{1/2}$ and donicity for certain redox reactions in solution

The difference of the standard free enthalpies of two different solvates of a certain metal ion may be considered as due to the superimposition of electrostatic and nonelectrostatic contribution [124].

$$\Delta G^\circ = \Delta G^\circ_{el} + \Delta G^\circ_{n.el}$$

The electrostatic contribution is highly influenced by the dielectric constant of the medium; since no relationship is found between the dielectric constant and the half-wave potential (not even for ions with noble gas structure, such as $K^+$), it may again be concluded that covalent contributions should not be neglected in any solvation phenomena.

## 7.3. The Influence of Anions

The redox equilibria can be considerably shifted by the presence of additional donor units. Thus the redox potential in a donor solvent will be influenced by the presence of anions and it may be different for a metal chloride and a metal iodide. The effect becomes more pronounced if the supporting electrolyte contains anions which have donor properties. Such donor anions will compete with solvent molecules for coordination.

The results obtained are in accordance with the relative donor properties of the anions as discussed in the previous sections. Iodide ions exert a small influence in weak donor solvents and are practically without any action in strong donor solvents. Chloride ions cause considerable shifts in half-wave potentials the amount of which is increasing with decreasing donicity of the medium [118]. There is no change in $E_{1/2}$ in DMSO containing perchlorate or chloride ion since DMSO and the chloride ion have similar donor properties. Azide ions give a slight change in $E_{1/2}$ in DMSO, which again increases in a solvent of lower donicity. As expected the effect of azide ions is more pronounced than that of chloride ions (Fig. 28). The shift in $E_{1/2}$ may be estimated by comparing the differences in donicities between solvent molecules and added donor.

## 7.4. The Influence of Water

The presence of water may have an appreciable effect on $E_{1/2}$, since water is a fairly strong donor. It is known that it is extremely difficult to remove the last traces of water from any solvent and it is therefore of interest to know the influence of water. It is apparent that in solution of a strong donor such as DMF, DMA, DMSO or HMPA the presence of small amounts of water is not reflected in a shift of the half-wave potential. On the other hand, the half-wave potential is shifted to negative potential values by the presence of water in a weak donor solvent.

Thus for analytical applications a solvent of high donicity is to be preferred, since the rather elaborate removal of small amounts of water is not required and admission of water from the atmosphere may be tolerable.

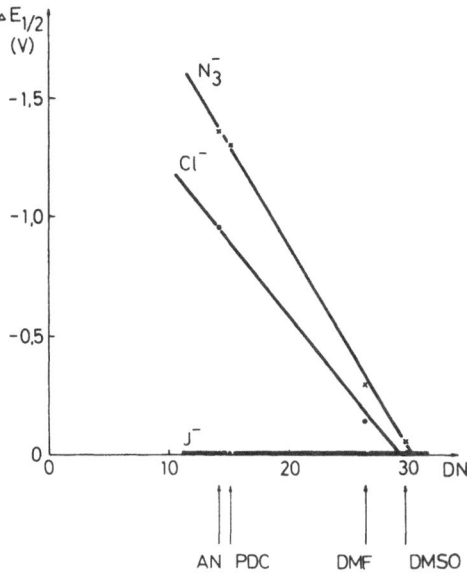

Fig. 28. $E_{1/2}$ for Eu(III)-Eu(II) in the presence of complex forming anions

It is obvious that this consideration concerning the presence of small amounts of water is not restricted to redox-equilibria and that it will also be important for ionization and complex equilibria in solvents of low donicity.

## 8. Conclusion

In the absence of thermodynamic data for the great variety of chemical interactions, particularly in non-aqueous solutions, the concept of donicity has been found an extremely useful guide for the
- (a) formation of donor-acceptor complexes,
- (b) ionization of covalent compounds by means of donor molecules,
- (c) autocomplex formation,
- (d) rate of solvent substitution,
- (e) solvation of metal ions and
- (f) redox-equilibria in non-aqueous media.

It is apparent that the exclusive application of the donicity-rule may fail to account for the actual reaction which is occurring, since solvation of anions, steric factors, and other specific donor-acceptor interactions as well as the entropic effects must also be considered.

However, the present outline should have illustrated, that in a notable number of coordination chemistry phenomena as well as ionization- and redox equilibria the concept of donicity serves as a usefule guide to the selection of a suitable medium for a particular purpose and to the prediction − at least qualitatively − of the properties of a given solution.

## 9. References

1) Davies, E. W.: Ion Association. London: Butterworth 1962.
2) Gurney, R. W.: Ionic Processes in Solution. London-New York: McGraw Hill 1961.
3) Drago, R. S., Purcell, K. F., in: Non-Aqueous Solvent Systems (ed. Waddington, T. C.). London-New York: Academic Press 1965.
4) Gutmann, V.: Coordination Chemistry in Non-Aqueous Solutions. Wien-New York: Springer 1968.
5) Chem. Zeit 4, 90 (1970). Angew. Chem. Int. Ed. 9, 843 (1970).
6) Briegleb, G.: Elektronen-Donator-Akzeptor-Komplexe. Berlin-Göttingen-Heidelberg: Springer 1961.
7) Gutmann, V., Wychera, E.: Inorg. Nucl. Chem. Letters 2, 257 (1966).
8) Spaziante, P., Gutmann, V.: Inorg. Chim. Acta, 5, 273 (1971). − Gutmann, V.: Rev. Chim. Min. 8, 429 (1971).
9) Gutmann, V., Czuba, H.: Monatsh. Chem. 100, 788 (1969).
10) − Mayer, U.: Monatsh. Chem. 100, 2048 (1969).
11) − Chimia 23, 285 (1969).
12) Mayer, U., Gutmann, V.: Monatsh. Chem. 101, 912 (1970).
13) Gutmann, V.: Rec. Chem. Progr. 30, 169 (1969).
14) Ahrland, S., Chatt, J., Davies, N. R.: Quart. Rev. (London) 12, 265 (1958).
15) Pearson, R. G.: J. Am. Chem. Soc. 85, 3533 (1963).
16) Beattie, I. R., Leigh, G. L.: J. Inorg. Nucl. Chem. 23, 55 (1961).
17) Tanaka, T., et al.: Inorg. Chim. Acta 3, 187 (1969).
18) Wannagat, U., Schwarz, R.: Z. Anorg. Allgem. Chem. 277, 73 (1954).
19) Vielberg, F.: Z. Anorg. Allgem. Chem. 291, 310 (1957).
20) Muetterties, E. L.: J. Inorg. Nucl. Chem. 15, 182 (1960).
21) Beattie, I. R., Gilson, T., Webster, M., McQuillan, G. P.: J. Chem. Soc. 1964, 238.
22) Gutmann, V., Mayer, U.: Monatsh. Chem. 99, 1383 (1968).
23) Wertz, D. J., Kruh, R. F.: Inorg. Chem. 9, 595 (1970).
24) Chambers, R. D., Mobbs, R. H., in: Advances in Fluorine Chemistry, Vol.4 (eds. Stacey, M., Tatlow, J. C., Sharpe, A. G.). London: Butterworth 1965.
25) Miller, W. T., Fried, J. H., Goldwhite, H. J.: J. Am. Chem. Soc. 82, 3091 (1960).
26) Graham, D.P., Weinmayr, V., McCormack, W. B.: J. Org. Chem. 31, 955 (1966).
27) West, R., Sado, A., Tobey, S.W.: J. Am. Chem. Soc. 88, 2488 (1966); Accounts Chem. Res. 3, 130 (1970).
28) Ebel, H. F., Schneider, R.: Angew. Chem. 77, 914 (1965).
29) Powell, J., Shaw, B. L.: J. Chem. Soc.(A) 1968, 774.
30) Nyholm, R. S.: Colloque Int. sur la Nature et les Proprieties des Liaisons de Coordination, Paris 1969.

31) Fischer, E. O., Oefele, K.: Z. Naturforsch. *14b*, 763 (1959); Chem. Ber. *93*, 1156 (1960).
32) Abel, E. W., Bennett, M. A., Wilkinson, G.: Chem. Ind. (London) *1960*, 442. – Abel, E. W., Butler, I. S., Reid, J. G.: J. Chem. Soc. *1963*, 2068.
33) Cassar, L., Foa, M.: Inorg. Nucl. Chem. Letters *6*, 291 (1970).
34) Hetnarski, B., Graboswki, Z., Kutkiewicz, W.: Roczniki Chem. *43*, 1589 (1969).
35) Muetterties, E. L.: J. Inorg. Nucl. Chem. *15*, 182 (1960).
36) Ryschkewitsch, G. E., Zutshi, K.: Inorg. Chem. *9*, 411 (1970).
37) Pitochelli, A. R., Ettinger, R., Dupont, J. A., Hawthorne, M. F.: J. Am. Chem. Soc. *84*, 1057 (1962). – Achran, G. E., Shore, S. G.: Inorg. Chem. *4*, 125 (1965).
38) Ryschkewitsch, G. E., Mathur, M. A., Sullivan, T. E.: Chem. Commun. *1970*, 117.
39) Benkeser, R. A., Foley, K. M., Grutzner, J. B., Smith, E. W.: J. Am. Chem. Soc. *92*, 697 (1970). – Bernstein, S. C.: J. Am. Chem. Soc. *92*, 699 (1970).
40) Campbell-Ferguson, H. J., Ebsworth, E. A. V.: Chem. Ind. (London) *1965*, 301.
41) Corey, J. Y., West, R.: J. Am. Chem. Soc. *85*, 4034 (1963).
42) Aylett, B. J., Campbell, J. M.: J. Chem. Soc. (A) *1969*, 1920.
43) Gutmann, V., Bohunovsky, O.: Monatsh. Chem. *99*, 740 (1968).
44) – Bardy, H.: Monatsh. Chem. *99*, 763 (1968).
45) – Laussegger, H.: Monatsh. Chem. *99*, 963 (1968).
46) Fowles, G. W. A., Lester, T. E.: J. Chem. Soc (A) *1968*, 1180.
47) Gutmann, V., Paulsen, G.: Monatsh. Chem. *100*, 358 (1969).
48) Cotton, F. A., Lippard, S. J.: J. Am. Chem. Soc. *88*, 1882 (1966).
49) Bagnall, K. W., Brown, D., Jones, P. J., du Preez, J. G. H.: J. Chem. Soc. *1966*, 737.
50) – – – Robinson, P. S.: J. Chem. Soc. *1964*, 2531.
51) – – – du Preez, J. G. H.: J. Chem. Soc. *1965*, 3594.
52) – Deane, A. M., Markin, T.C., Robinson, P. S., Stewart, M. A. A.: J. Chem. Soc. *1961*, 1611.
53) – Brown, D., Cotton, R.: J. Chem. Soc. *1964*, 2527.
54) Gaizer, F., Beck, M. T.: J. Inorg. Nucl. Chem. *29*, 21 (1967).
55) Cotton, F. A., Francis, R.: J. Am. Chem. Soc. *82*, 2986 (1960).
56) Gutmann, V., Schmid, R.: Unpublished.
57) – Wegleitner, K. H.: Monatsh. Chem. *99*, 268 (1968). – Turco, A., Pecite, G., Niccolini, M.: J. Chem. Soc. *1962*, 3008.
58) Buffagni, S., Dunn, T. M.: J. Chem. Soc. *1961*, 5105.
59 Gutmann, V., Hampel, G., Masaguer, J. R.: Monatsh. Chem. *94*, 822 (1963).
60) Janz, G. J., Marcinkovsky, A. E., Venkatasetty, H. V.: Electrochim. Acta *8*, 867 (1963).
61) Baaz, M., Gutmann, V., Hampel, G., Masaguer, J. R.: Monatsh. Chem. *93*, 1416 (1962).
62) Libus, W.: Roczniki Chem. *36*, 999 (1962).
63) Roczniki Chem. *35*, 411 (1961).
64) Gutmann, V., Fenkart, K.: Monats. Chem. *98*, 1 (1967).
65) – Bohunowsky, O.; Monatsh. Chem. *99*, 751 (1968).
66) – Leitmann, O.: Monatsh. Chem. *97*, 926 (1966).
67) – Scherhaufer, A.: Monatsh. Chem. *99*, 1686 (1968).
68) – – Inorg. Chim. Acta *2*, 325 (1968).
69) Fine, D. A.: J. Am. Chem. Soc. *84*, 1139 (1962).
70) Katzin, L. I., Gebert, E.: J. Am. Chem. Soc. *72*, 5464 (1950).
71) – – J. Am. Chem. Soc. *72*, 5659 (1950).
72) Senise, P.: J. Am. Chem. Soc. *81*, 4196 (1959).

73) Babko, A. K., Drako, O. F.: J. Gen. Chem. (USSR) *19*, 1809 (1949).
74) Gutmann, V., Mayer, U.: Unpublished.
75) Lehne, M.: Bull. Soc. Chim. France *1951*, 76.
76) Bobtelsky, M., Spiegler, K. S.: J. Chem. Soc. *1949*, 143.
77) Gutmann, V., Hübner, L.: Monatsh. Chem. *92*, 1261 (1961).
78) − Weisz, A.: Monatsh. Chem. *100*, 2104 (1969).
79) Weisz, A., Gutmann, V.: Monatsh. Chem. *101*, 19 (1970).
80) Egghart, H. C.: J. Inorg. Nucl. Chem. *31*, 1538 (1969).
81) Jørgensen, C. K.: Inorganic Complexes, p. 42. London-New-York: Academic Press 1963.
82) Gutmann, V., Bardy, H.: Z. Anorg. Allgem. Chem. *361*, 213 (1968).
83) − Weisz, A., Kerber, W.: Monatsh. Chem. *100*, 2096 (1969).
84) − Fenkart, K.: Monatsh. Chem. *99*, 1452 (1968).
85) − Beer, G.: Inorg. Chim. Acta *3*, 87 (1969).
86) Peterson, R. J., Lingane, P. J., Reynolds, W. L.: Inorg. Chem. *9*, 680 (1970).
87) Buckingham, A., Gasser, R. P. H.: J. Chem. Soc. (A) *1967*, 1964.
88) Mayer, U., Gutmann, V.: Monatsh. Chem., in the press.
89) Gutmann, V., Wegleitner, K. H.: Monatsh. Chem. *101*, 1532 (1970).
90) − Imhof, J.: Monatsh. Chem. *101*, 1 (1970).
91) Sutton, G. J.: Australian J. Chem. *11*, 415, 420 (1958).
92) Roper, W. R., Wilkins, C. J.: Inorg. Chem. *3*, 500 (1964).
93) Deveny, M. J., Webster, M.: J. Chem. Soc (A) *1970*, 1643.
94) Chini, P.: Inorg. Chim. Acta Rev. *2*, 48 (1968).
95) Hieber, W., Schubert, E. H.: Z. Anorg. Allgem. Chem. *338*, 37 (1965).
96) − Brendel, G.: Z. Anorg. Allgem. Chem. *289*, 338 (1957).
97) − Werner, R.: Chem. Ber. *90*, 286, 1116 (1957).
98) − Sedlmeier, J., Werner, R.: Chem. Ber. *90*, 278 (1957).
99) − Lipp, A.: Chem. Ber. *92*, 2085 (1959).
100) Heck, R. F.: J. Am. Chem. Soc. *85*, 657 (1963).
101) Gutmann, V., Schmid, R.: Monatsh. Chem., in the press.
102) − − Unpublished.
103) Baaz, M., Gutmann, V., Kunze, O.: Monatsh. Chem. *93*, 1162 (1962).
104) Eigen, M., Maas, G.: Z. Physik Chem. N. F. *49*, 163 (1966).
105) Czerlinski, G., Diebler, H., Eigen, M.: Z. Physik Chem. N. F. *19*, 246 (1959).
106) Eigen, M., Tamm, K.: Z. Elektrochem. *66*, 93, 107 (1962).
107) − Z. Elektrochem. *64*, 115 (1960).
108) − Wilkins, R. G.: Advan. Chem. *49*, 55 (1965).
109) − Pure Appl. Chem. *6*, 97 (1963).
110) Basolo, F., Pearson, R. G.: Mechanisms of Inorganic Reactions, p. 88. New York: John Wiley and Sons Inc. 1967.
111) Eigen, M.: Proc. VII. I. C. C. C., Stockholm 1962.
112) Erlich, R. H., Roach, E., A. I.: Popov, private communication.
113) Gutmann, V., Schöber, G.: Monatsh. Chem. *90*, 897 (1959).
114) Rusina, A., Schroer, H. P.: Collection Czech. Chem. Commun. *31*, 2600 (1966).
115) Schroer, H. P., Vlček, A. A.: Z. Anorg. Allgem. Chem. *334*, 205 (1964).
116) Vlček, A. A.: Z. Anorg. Allgem. Chem. *304*, 109 (1960).
117) Gritzner, G., Gutmann, V., Schmid, R.: Electrochim. Acta *13*, 919 (1968).
118) Gutmann, V., Peychal-Heiling, G.: Monatsh. Chem. *100*, 1423 (1969).
119) − − Michlmayr, M.: Inorg. Nucl. Chem. Letters *3*, 501 (1967).
120) − Allgem. Prakt. Chem. *21*, 116 (1970).

[121]  – Schmid, R.: Monatsh. Chem. *100*, 2113 (1969).
[122]  Jørgensen, C. K.: Inorganic Complexes, p. 21 ff. London-New York: Academic Press 1963.
[123]  Katzin, L. I.: J. Chem. Phys. *23*, 2055 (1955); *36*, 3034 (1962).
[124]  Kolthoff, I. M.: J. Polymer Sci. *10*, 22 (1964).

Received September 3, 1970

# Solvent Effects and NMR Coupling Constants

## Prof. Dr. Stanford L. Smith

University of Kentucky, Department of Chemistry, Lexington, Kentucky, USA

## Contents

# I. Introduction

*Early in the development of NMR techniques it was recognized that chemical shifts varied* with the physical state of the sample and in the case of liquid samples *with the nature of the solvent.* Coupling constants were believed to be invariant to changes of solvent except where gross changes in bonding or molecular conformation occurred. During the late 1950's and early 1960's isolated cases of what might be solvent dependent coupling constants appeared in publications basically devoted to other studies. In 1963 and 1964 it became apparent that coupling constants were not invariant to solvent. Since that time over three dozen different coupling constants involving a dozen different nuclei have been shown to vary with the solvent in which the sample is observed. It is the purpose of this article to review these developments.

Obviously, a discussion of solvent effects and coupling constants requires some discussion of solvent effects on chemical shifts and opens the way for a general discussion of solute-solvent interactions. In virtually every instance solvent induced changes in coupling constants parallel the solvent induced changes in chemical shifts (or vice versa). The only exceptions (and those appear to be universal) are aromatic solvents which give rise to chemical shift changes by virtue of diamagnetic anisotropy effects which are not accompanied by corresponding changes in the coupling constants. Two excellent reviews [1,2] of solvent effects and NMR parameters have appeared within the last few years. This article will be restricted as far as possible to the specific question of coupling constants and solvent effects. However, our discussions will be facilitated by a brief consideration of the origin of nuclear spin-spin coupling constants, the nature of solvent-solute interactions, and the general mechanisms by which solvent-solute interactions are thought to affect coupling constants.

Following these introductory considerations solvent dependent coupling constants will be reviewed in order of the number of bonds intervening between the coupled nuclei. This organization does not follow the historical development of the work in this area nor does it necessarily follow the major developments or principles proposed. It does provide a convenient way of organizing an otherwise unwieldly body of information.

# II. Background

## A. Origin of Nuclear Spin-Spin Coupling

The theory of spin-spin coupling has been developed extensively in a number of works such as that of Emsley, Feeney and Sutcliffe [3] to which the reader is referred for detailed treatments. For our purposes the nuclear spin-spin coupling constant $J_{NN'}$ can be written as the sum of four contributions

$$J_{NN'} = J_{NN'}^{(1)} + J_{NN'}^{(2)} + J_{NN'}^{(3)} + J_{NN'}^{(4)} .$$

The last term represents the internuclear dipole-dipole interaction which gives rise to the broad lines observed in solids. It averages to zero when all molecular orientations are equally probable and is probably not important for liquid samples.

The first three terms represent different contributions to the nucleus-electron ... electron-nucleus interaction. The first term, $J_{NN'}^{(1)}$, represents the interaction between the nuclear magnetic moment and the *orbital* magnetic moments of the electrons, the *spin-orbit* interaction. The second term, $J_{NN'}^{(2)}$, represents the interaction between the nuclear magnetic dipole and the electron magnetic dipole, the *dipole-dipole* interaction. Both of these interactions depend on the electron distribution and orbital energies (both ground state and excited state) of the molecule under consideration. Both contributions have been shown to be small ($< 7\%$) for couplings involving hydrogen and at least minor for couplings involving other nuclei [3]. As a result, with a few exceptions [4] relatively little has been done to develop these quantities in terms of approximate molecular orbital or valance-bond calculations as applied to reasonably complicated molecules.

The third and most important term, $J_{NN'}^{(3)}$, depends on the properties of electrons *at the nucleus* and hence is known as the contact term. Since only *s* orbitals have appreciable magnitude at the nucleus the development of this term requires the evaluation of the magnitude of *s* functions for various electronic models of a molecule. Because it is the major contributor to indirect coupling for hydrogen and probably for most other nuclei the contact term has been developed in terms of approximate methods of calculating electronic structures of molecules. For our purpose the MO theory development by Pople and Santry [4] and by Pople and Bothner-By [5] is useful. Using the contact contribution the coupling constant is given by

$$J_{AB} = K S_A^2(0) S_B^2(0) \sum_i^{occ} \sum_j^{unocc} (\epsilon_j - \epsilon_i)^{-1} C_{iA} C_{iB} C_{jA} C_{jB} \qquad (1)$$

$$K = \frac{64}{9} h \beta^2 \gamma_A \gamma_B$$

where $(\epsilon_j - \epsilon_i)$ is the energy difference between an occupied and an unoccupied molecular orbital, $S_N^2$ is the probability density of a valence *s* orbital at the nucleus (A or B), and $C_{iN}$ and $C_{jN}$ are the coefficients of the atomic orbital $S_N$ in the *i*th and *j*th MOs. In practice, further simplification is obtained by making an average energy approximation, thus removing the need to evaluate each of the excitation energies in the summations. This model has produced excellent qualitative and in some cases quantitative explanations of trends in coupling constants as a function of the atoms involved and of substituent effects on couplings. We shall use it later to rationalize the mechanism of solvent effects on coupling constants.

We will be concerned later with the fact that coupling constants may be either positive or negative. A positive coupling reflects the fact that interaction between nuclei whose spins are opposed results in a more stable system, and a negative coupling reflects the opposite situation, *so long as the magnetogyric ratios of the coupled nuclei are of the same sign.* In order to correct for the effect in Eq. (1) when $\gamma_A$ and $\gamma_B$ have opposite signs it will be convenient to consider the reduced coupling constants; $J_{AB}/\gamma_A\gamma_B$. This effectively isolates the electronic contributions and effects from the nuclear effects. It will further be convenient to consider only the *sign* of the reduced coupling rather than subject the reader to numerous conversions in units and magnitudes from those commonly in use.

## B. Solvent-Solute Interactions

Interactions between a solute and a solvent may be broadly divided into three types; specific interactions, reaction field and Stark effects, and London-van-der-Waals or dispersion interactions. Specific interactions involve such phenomena as ion pair formation, hydrogen bonding and $\pi$-complexing. Reaction field effects involve the polarization of the surrounding nonpolar solvent by a polar solute molecule resulting in a solvent electric field at the solute molecule. Stark effects involve the polarization of a non-polar solute by polar solvent molecules. Dispersion interactions, generally the weakest of the three types, involves nonpolar solutes and nonpolar solvents via snap-shot dipole interactions, etc. For our purposes it is necessary to develop both the qualitative and semiquantitative forms in which these kinds of interactions are encountered in studies of solvent effects on coupling constants.

Considerations of *specific interactions* requires that we distinguish between the strength of various types of interactions. Very strong interactions, e.g., salt formation from a basic compound dissolved in an acidic solvent, effectively result in the formation of a new compound whose bonding and molecular conformation are quite different from those of the parent substance. Generally, such strong interactions are not considered as solvent effects in the context we are considering. Weaker interactions involving energies in the order of a few kilocalories or less are the primary interest.

The most common weak specific interaction is hydrogen bonding. More often than not hydrogen bonding is invoked by various investigators on the basis of analogy, chemical intuition, knowledge of solvent molecular structure, or because the data do not correlate well with other models. In a few cases substantive data such as variations of bond stretching frequencies in infrared spectra are reported. A few examples of collision complex models or of equilibrium or thermodynamic models based on chemical shift variations have been used. In no case have quantitative measurements such as $pK_A$'s, heats of solution and the like been correlated with solvent effects on coupling constants. In part this situation seems to arise because such correlations have not been clearly needed. Also, it is frequently the case that either the solvents employed, the solutes investigated or both have not been sufficiently well studied to provide the necessary data for such approaches.

Another specific interaction involves solute dipole-solvent dipole pairing. This is usually proposed when the change in the coupling constant seems to be linearily related to the dipole moments of the solvents.

Pi-complexing is most commonly used to rationalize effects observed in aromatic solvents. The most frequent evidence cited is magnetic anisotropy effects on chemical shifts in the solute molecule. As was the case for hydrogen bonding no quantitative correlations with substantive parameters such as ultraviolet spectral shifts have been attempted.

With a few exceptions solvent dependent coupling constants have been observed only in non-ionic compounds. As a result no data are available concerning correlations of coupling constant changes and heats of solvation, heats of solution, ion pairing, etc.

*Reaction field* interactions have received by far the most attention in studies of solvent dependent couplings. The secondary electric field induced in the polar (or polarizable) solvent by the permanent dipole moment of the solute, in turn affects the electronic structure of the solute molecule to an extent dependent on the polarizability of the solute molecule. Onsager [6] developed the original model of a point dipole solute at the center of a spherical cavity in a homogeneous, continuous, polarizable medium of dielectric constant $\epsilon$. Qualitatively this model requires that the magnitude of the electric field at the solute molecule depend on the solute dipole moment $\mu$, the solute polarizability $\alpha$, the solvent dielectric constant $\epsilon$, and the shape of the solute cavity. The model also requires that the reaction field be parallel to the solute dipole moment. Correlations between $\epsilon$, $\epsilon^{1/2}$ or $\epsilon^2$ and changes in coupling constants are frequently presented as qualitative evidence for some sort of reaction field effect on coupling constants.

The quantitative expression for the reaction field $R$ is

$$R = \frac{\mu(n^2 - 1)}{3\alpha} \frac{(\epsilon - 1)}{(\epsilon + n^2/2)} \tag{2}$$

where the polarizability $\alpha$ is

$$\alpha = \frac{(n^2 - 1) r^3}{(n^2 + 2)}$$

and $n$ is the refractive index of the *solute*. Diehl and Freeman [7] have extended this model (as applied to chemical shift effects) to the case of ellipsoidal cavities, but this extension has not been used in studies of coupling constants. Buckingham *et al.* [8] have shown that, to a good approximation, the shielding of a nucleus in a chemical bond should be a linear function of the field component along that bond

$$\underset{R}{\Delta\sigma} = k_R \cdot R \cdot \cos\theta \tag{3}$$

where $R \cdot \cos\theta$ is the field component along the bond in question and $k_R$ is a constant characteristic of the bond in question. For the reaction field model $\cos\theta$ is the angle between the net solute dipole and the bond being studied. The same kind of linear relationship is assumed implicitly in all discussions of solvent effects on coupling constants.

As a result of the (assumed) linear relationship with reaction field and the quantitative expression for $R$ numerous investigators report correlations or the lack thereof with expressions such as

$$\frac{(\epsilon - 1)}{2\epsilon + 2.5}, \quad \frac{(\epsilon - 1)}{(\epsilon + 1)}, \quad \frac{(\epsilon - 1)}{(2\epsilon + n^2)}. \tag{4}$$

In all cases the dielectric constant used is that of the pure solvent. Neglect of the solute is usually justified by its low concentration and the assumption that any necessary correction would be additive. In at least a few cases where the first two expressions have been employed the linearity of the results is to some extent dependent on how closely the refractive index of the solute meets the conditions $n^2 = 2.0$ or $2.5$; a situation not always recognized by the investigators. In one instance attempts have been made to clarify the role of solvent reaction field by examining solutes with different dipole moment orientations relative to the bonds involving coupled atoms.

The solvent Stark term developed by Baur and Nicols [9] reflects the same qualitative interactions as the reaction field term, however, it concerns the situation when the solute is less polar (in the ideal case non-polar) than the surrounding solvent. Correlations with Stark effects are usually recognized as linear relations to the term

$$(\epsilon - 1)(2\epsilon + 1)/\epsilon$$

which in the limit of large values of the dielectric constant approximates $\epsilon$. Unlike the reaction field term the Stark term does not depend on the existence of a solute dipole moment nor on its orientation if the dipole does exist. Relatively few investigators concerned with the solvent effects on coupling constants have considered the Stark term. In a few cases investigators have noted that some solute parameter shows reasonable correlations with the reaction field term for solvents having dielectric constants smaller than that of the solute and correla-

123

tions with the Stark term for solvents whose dielectric constant is larger than that of the solute.

*Dispersion interactions* have been shown in the absence of other effects to be responsible for gas-to-liquid changes of chemical shifts [1,2]. The dispersion contribution to the electric field effect on infrared and ultraviolet spectral transitions has been shown to be proportional to McRae term [10,11]

$$\frac{n^2 - 1}{2n^2 + 1} \tag{5}$$

where $n$ is the refractive index of the *solvent*. Correlations between the change of coupling constants in centrosymmetric (nonpolar) molecules and the McRae term are taken as evidence that dispersion interactions are involved. An alternate approach is to assume a linear relationship between the heat of vaporization of a nonpolar solvent at its boiling point and the strength of the dispersion interactions between the solvent molecules [8]. A more qualitative view is provided by recognizing that dispersion interactions are expected to be greater for molecules having heavier atoms, e.g. halogens, or by seeking correlations with $n$, $n^{1/2}$, $n^2$, etc.

## C. Mechanism of Solvent Effects on Coupling Constants

It is necessary to make a clear distinction between the mechanism by which solute molecules *interact* with solvent molecules, and the mechanism by which the solvent *affects a change* in a coupling constant. A variety of interaction mechanisms are conceivable and evidence for most of them has been found. The important point is that changes in a coupling constant must reflect changes in the electronic structure of the solute molecule (assuming as we have that direct nuclear dipole-nuclear dipole contributions to the coupling are zero for liquid samples). The only ways to change the electronic structure of a ground state solute molecule are to change the time average equilibrium positions of the atomic nuclei or to subject the molecule to external magnetic or electric fields (or both). Certainly strong interactions or extreme changes in temperature might accomplish the former effect. For the weak interactions we are considering, magnetic or electric fields seem the most likely causative factors.

All of the interaction mechanisms described above are expected to produce electric fields in the solute cavity. In the case of specific interactions and reaction field effects these electric fields are expected to have some specific orientation with respect to the solute coordinate system. Dispersion forces and Stark effects are not expected to have any specific orientation with respect to the solute. Magnetic field effects seem unlikely to be important in light of the well-known invariance of coupling constants to changes of the external magnetic field. However, it is conceivable that a solvent "magnetic reaction field" might exist where the magnetic moment of one nucleus induces a magnetic polarization of the surrounding medium which in turn affects another nucleus. Raynes [12] gives the expression for such a contribution as

$$J_m = -\frac{8\pi}{3}\frac{\chi_v}{ha^3}h^2\,\gamma_A\gamma_B \tag{6}$$

where $\chi_v$ is the volume magnetic susceptability of the solution, $\gamma_A$ and $\gamma_B$ are the magnetogyric ratios of the coupled nuclei and $a$ is the radius of a spherical cavity containing the solute molecule. For coupled protons of a solute dissolved in $CCl_4$ with $a = 2\text{Å}$ he calculates $J_m = 0.1$ Hz. Thus, it appears that magnetic field effects are at least small if not completely absent.

The idea that an electric field component directed along the bond(s) connecting coupled nuclei would cause changes in electron distribution resulting in changes in coupling constants was first introduced by Smith and Cox [13] and by Bell and Danyluk [14]. Their rationalizations were developed primarily to explain the relationship between the absolute signs of coupling constants and the observed changes in the couplings as a function of solvent. These arguments will be considered in more detail later.

Raynes and Sutherley [15] have presented the most definitive evidence for electric field effects on coupling constants and have shown that the experimental data is in good agreement with the theoretically predicted effect. They observed $^1J_{C\text{-}H}$ in dimethylformamide (DMF) and in the complex $[Al(DMF)_6]^{+3}$. DMF is known to coordinate through the oxygen atom. Owing to hindered rotation around the C-N bond the two methyl groups give well resolved signals separated by 0.18 p.p.m. at 30 °C. Upon coordination with $Al^{+3}$ both methyl signals move downfield by 0.28 p.p.m. while the aldehyde proton moves downfield by 0.30 p.p.m. It is assumed that the downfield shift of the methyl signals arises primarily from a "through space" electric field. The aldehyde proton also experiences this "through space" electric field, but probably also experiences some noticeable inductive effect and it is difficult to separate the two effects. Buckingham's equation (Eq. 3) provides a good estimate of the electric field at the methyl groups. Knowing the electric field a measurement of the change in $^1J_{C\text{-}H}$ of the methyl groups in going from "free" to "bound" DMF permits an evaluation of the electric field dependance of $^1J_{CH}$ for DMF. For 0.3M $Al(NO_3)_3$ in DMF at 30 °C $^1J_{C\text{-}H}$ increases from 137 Hz to 143 Hz on going from the free to the bound state. This leads to

$$\frac{dJ}{dE_z} = +24 \times 10^{-6} \text{ Hz (Field in e.s.u.)}^{-1},$$

where the axis of the methyl group is taken as the $z$ axis and the field is directed from the carbon atom to the plane of the three protons. Essentially the same value was obtained using data for $^1J_{C-H}$ in a variety of complexes of the form M(acetylacetone)$_n$ [16].

A theoretical value for the magnitude of $dJ/dE_z$ was obtained using the delocalized molecular orbital approach of Gil and Teixeira-Dias [17] who calculated substituent effects on $^1J_{CH}$. The Pople expression for the contact contribution to the coupling constant $^1J_{C-H_1}$ of a methyl group can be written

$$J_{C-H} = C\pi_{sh_1} \tag{7}$$

where $C$ is a constant and $\pi_{sh_1}$ (equal to the double summation of Eq. 1) is the atom-atom polarizability of the hydrogen $1s$ orbital and the carbon $2s$ orbital. The electric field leads to a change in $\pi_{sh_1}$ because of

(a) a change in the resonance integral $\beta_{sh_1}$, and

(b) changes in the coulomb integrals $\alpha_{h_2}$, $\alpha_{h_3}$ and $\alpha_{h_4}$.

The contribution from changes in the resonance integral is calculated to be

$$\frac{dJ}{dE_z}^{(a)} = +10 \times 10^{-6} \, \text{Hz (field in e.s.u.)}^{-1}.$$

The effect of changes in the coulomb integrals is shown to be

$$\frac{dJ}{dE_z}^{(b)} = +6 \times 10^{-6} \, \text{Hz (field in e.s.u.)}^{-1}.$$

Other contributions are shown to have opposite signs and cancel or to be very small. The total calculated effect

$$\frac{dJ}{dE_z} = +16 \times 10^{-6} \, \text{(field in e.s.u.)}^{-1}$$

is in excellent agreement with the experimentally determined values. The observed value may be an upper limit for the electric field effect since it undoubtedly includes not only some small inductive effect, but also some small unspecified contribution from ion pairing. Using the above values Raynes suggests that for a small organic molecule having a methyl group and a dipole moment of 1 Debye, a simple reaction field model predicts changes in $^1J_{C-H}$ in the range of 0.2 to 0.5 Hz for solution in "inert", non-polar solvents. These values fall towards the lower end of the ranges of changes which have actually been observed.

Obviously, complex formation with a metal ion is a strong interaction rather than the weak interactions of primary interest here. However, the Raynes experiment and calculations clearly establish both the existence and theoretical justification for electric field effects on coupling constants. It seems reasonable to assume that the general mechanism operates for weak solute-solvent interactions. Raynes [12] has further suggested an empirical partitioning of the change,

$\Delta J$, of a coupling constant in going from gas phase at low pressure to solution. Thus,

$$\Delta J = J_m + J_w + J_E + J_c \qquad (8)$$

where $J_m$ is the magnetic reaction field contribution discussed previously (Eq. 7), $J_w$ refers to the change brought about by dispersion interactions, $J_E$ denotes the effect of solvent electric fields (reaction fields) and $J_c$ reflects the effect of specific interactions. Unfortunately, gas phase values of the coupling constants studied by most authors are not available. This approach can still be applied if values of the coupling constant in non-interacting, low dielectric constant, low refractive index solvents such as neopentane, tetramethylsilane or hexane are substituted for the values of the coupling constant at low pressure in the gas phase. The numerical values obtained for various contributions will be in error, but the relative magnitudes of the different effects can be estimated. With few exceptions no attempts have been made to evaluate combined contributions of different interaction mechanisms to the change observed for a given coupling constant. Frequently, this is because the system studied was selected to optimize the study of a particular interaction or to facilitate the examination of a particular coupling.

Some investigations have proposed mechanisms for the effect of solvents on couplings constants which appear to be different from the electric field approach presented above. This dicotomy needs clarification. As noted earlier the problem is to distinguish between the solute-solvent interaction mechanism and the forces which ultimately result in a change in a coupling constant. In part the problem is one of the language used. It is convenient and correct to say that two (or more) molecules interact by hydrogen bonding. A logical extension is to state that the observed coupling constant changes because of this hydrogen bonding; which may also be correct in so far as it goes. The difficulty is that it doesn't go far enough. Even extending the explanation by saying that increased hydrogen bonding leads to a lengthening of some bond (which is not necessarily true) which results in a change in the coupling constant is not sufficient. An electron-rich atom, e.g., oxygen or nitrogen, involved in a hydrogen bond constitutes a center of charge which results in an electric field just as does the $Al^{+3}$ ion in the experiment described above. As a result of the interaction with the hydrogen that electric field will have some specific orientation with respect to the bonds in the solute molecule, just as does the field of the $Al^{+3}$ ion. Thus, in so far as hydrogen bonding is considered to be a weak electrostatic interaction the fundamental mechanism changing the bonds in question and as a result changing the coupling constant is an electric field effect. Only in the case where there is actual bonding, i.e., significant orbital overlap and electron exchange, can we consider the possibility of a true hydrogen *bonding* effect on a coupling constant. However, such strong interactions exceed the limits of the weak solute-solvent interactions with which we are concerned here.

127

In a similar fashion we may describe the solute-solvent interaction as a dispersion interaction and say a coupling constant changes because of dispersion effects. Again this is true in so far as it goes. However, dispersion interactions arise from dipole-induced dipole, induced dipole-induced dipole, collision induced time dependent molecular distortions, electron repulsion and correlation, etc. All of which effectively produce electric fields at the solute molecule.

It is inconvenient at the very least to attempt to calculate the electric and perhaps the magnetic fields arising from a particular interaction mechanism. Thus, we will continue to speak of *interaction* mechanisms and the *resulting effect* on coupling constants as ' . . . the change arising from dispersion interactions . . . ' or 'The reaction field effect on coupling . . .'. It will be useful to seek empirical relations between different interaction mechanisms and the resulting effects in the partitioned framework proposed by Raynes. Similarly, it will be usefull to explore the effect of solvents on different types of coupling constants, e.g., geminal H-H couplings, vicinal H-F couplings, etc., in terms of qualitative models such as resonance contributions, neighboring orbital contributions, degree of *s* character and the like, recognizing that these are limited but useful reflections of the electronic structure changes treated more fully in the theoretical approaches of Pople *et al.* and of Raynes *et al.*

## III. Solvent Dependent Coupling Constants

### A. General Comments

It is convenient to introduce the term $\Delta J$ to represent the difference in magnitude of a coupling constant measured in some low dielectric constant or low refractive index solvent compared to the magnitude of the same coupling constant measured in a solvent of high dielectric constant or refractive index. Where the absolute sign of a coupling constant is known the sign of $\Delta J$ indicates the direction of the observed change. This is a matter of empirical convenience since different investigators use different series of solvents and concentrations are not always extrapolated to infinite dilution. On occasion $\Delta J$ may simply represent the difference between the largest and smallest values measured for some coupling in different solvents. The value of this approach lies solely in providing the reader with a ready indication of the magnitude and direction of the changes observed.

In a similar fashion, it is noteworthy to report the solvents used in a given study using standard abreviations, e.g., TMS for tetramethylsilane, and list refractive indices and/or dielectric constants rather than calculated values for the reaction field or the McRae dispersion term.

## B. One-Bond Coupling Constants

### 1. $^1J_{^{13}C\text{-}H}$ and $^1J_{^{13}C\text{-}F}$

In one of the earliest papers devoted primarily to solvent effects on coupling constants Evans [18] reported a 9.6 Hz increase for $J_{C\text{-}H}$ of chloroform in thirteen solvents ranging from cyclohexane ($J = 208.1$ Hz) to dimethylsulfoxide ($J = 217.7$ Hz) (Table 1). Laszlo [19] provided additional values in a series of oxygen and nitrogen heterocyclic solvents (Table 2). Both authors attributed the changes in $^1J_{C\text{-}H}$ to hydrogen bonding, which is certainly reasonable and consistent with the data. However, the evidence is not overwhelming. Absolute measures for the basicity of the heterocycles in Table 2 are not available and there is some question whether or not ionization values given are for the hetero-atom lone pair. The trends suggested by that data are in opposite directions for the oxygen and nitrogen series.

Table 1. *$^{13}C$-H coupling constant of chloroform in various solvents*

| Solvent | $J_{^{13}C\text{-}H}$ * |
|---|---|
| Cyclohexane | 208.1 |
| Carbon tetrachloride | 208.4 |
| Chloroform | 209.5 |
| Benzene | 210.6 |
| Acetyl chloride | 211.8 |
| Nitromethane | 213.6 |
| Ether | 213.7 |
| Triethylamine | 214.2 |
| Methyl alcohol | 214.3 |
| Acetonitrile | 214.6 |
| Acetone | 215.2 |
| Dimethylformamide | 217.4 |
| Dimethylsulfoxide | 217.7 |

*Mole fraction of $CHCl_3$ = 0.15

The data in Table 1 provide a clearer picture. While the order of solvents for which $J_{C\text{-}H}$ increases is roughly the same as the order of increasing dielectric constants, there are several obvious discrepancies.

129

Table 2. $^{13}$C-H coupling constant of chloroform in nitrogen and oxygen heterocyclic solvents

| Solvent | Ionization potential | $J_{^{13}C\text{-}H*}$ |
|---|---|---|
| Pyridine | 9.76 | 215.0 ± 1.0 Hz |
| Pyridazine | 9.86 | 215.0 |
| s-Triazine | 10.07 | 211.0 |
| Pyrazine | 10.01 | 210.5 |
| Cyclohexane | (9.88) | 208.1 |
| Tetrahydrofuran | 9.54 | 214.0 |
| Dioxane | 9.13 | 213.0 |
| Dimethyl-2,6-γ-pyrone | ? | 212.0 |
| Paraldehyde | | 212.5 |
| Dihydropyran | 8.34 | 210.5 |

*Mole fraction of chloroform < 0.15

For example, nitromethane ($\epsilon \approx 35$) has about the same effect on $J_{C\text{-}H}$ as does diethyl ether ($\epsilon \approx 4.4$). Evans [18] develops the argument that hydrogen bonding might lengthen the C-H bond which ought to result in a decrease in $J_{^{13}C\text{-}H}$ or leave the bond length essentially unchanged, but via the electrostatic mechanism discussed in Sect. II. C increase the carbon 2s contribution resulting in an increase in $J_{^{13}C\text{-}H}$; the result actually observed. This argument is supported by the fact that the C-D stretching frequency of CDCl$_3$ is essentially the same in ether, acetone and inert solvents such as cyclohexane, and decreases only slightly in DMSO and DMF. The C-D stretching frequency does change markedly in triethylamine suggesting that the relatively low value of $J_{^{13}C\text{-}H}$ in that solvent is a result of competition between the decrease arising from lengthening of the C-H bond and the increase caused by the electrostatic repulsion mechanism.

Further support for this argument is given by the fact that phenylacetylene, a compound known to form hydrogen bonds roughly comparable in strength to those of chloroform, shows a change of only 1.2 Hz in $J_{^{13}C\text{-}H}$ between CCl$_4$ and DMSO. However, the C-H stretching frequency of phenylacetylene does show significant changes upon hydrogen bonding, again suggesting a competition between the electric field effect and the bond lengthening effect.

Additional evidence for the hydrogen bonding mechanism is provided by Douglas and Dietz [20] who observed that $J_{^{13}C\text{-}H}$ for chloroform increases with decreasing temperature as expected for a weak interaction such as hydrogen bonding. They note that the magnitude of the temperature effect ($0.7\,Hz/100\,°C$) is much too big to be accounted for by repartition of vibrational levels.

Watts and Goldstein [21] provide further support for the primacy of hydrogen bonding effects on $J_{13_{C-H}}$ by examining thirteen different halomethanes as the neat compound and as 20–25 mole % solutions in cyclohexane, $CCl_4$ and DMF (Table 3). The $\Delta J$ values they obtain increase with the proton-donating ability of the solute; $Cl > Br > I$.

Table 3. *Solvent effects on the* $J_{13_{C-H}}$ *of some halogenated methanes*

| Solute | Solvent | $J_{13_{C-H}}$(Hz) | $\Delta J$ (Hz) |
|---|---|---|---|
| | DMF | 216.46 | |
| $CHCl_3$ | | 208.91 | |
| | $CCl_4$ | 208.26 | 8.35 |
| | $C_6H_{12}$ | 208.11 | |
| | DMF | 211.60 | |
| $CHBr_3$ | | 205.40 | |
| | $CCl_4$ | 204.60 | 7.29 |
| | $C_6H_{12}$ | 204.31 | |
| | DMF | 180.55 | |
| $CH_2Cl_2$ | | 178.11 | |
| | $CCl_4$ | 176.75 | 4.07 |
| | $C_6H_{12}$ | 176.48 | |
| | DMF | 181.33 | |
| $CH_2BrCl$ | | 178.96 | |
| | $CCl_4$ | 177.70 | 3.95 |
| | $C_6H_{12}$ | 177.38 | |
| | DMF | 181.63 | |
| $CH_2Br_2$ | | 179.22 | |
| | $CCl_4$ | 177.98 | 3.89 |
| | $C_6H_{12}$ | 177.74 | |
| | DMF | 177.54 | |
| $CH_2BrI$ | | 176.20 | |
| | $CCl_4$ | 175.28 | 2.81 |
| | $C_6H_{12}$ | 174.73 | |

Table 3 (continued)

| Solute | Solvent | $J_{13_{C-H}}$ (Hz) | $\Delta J$ (Hz) |
|---|---|---|---|
| | DMF | 173.80 | |
| $CH_2I_2$ | | 172.92 | 1.87 |
| | $CCl_4$ | 172.15 | |
| | $C_6H_{12}$ | 171.93 | |
| | DMF | 162.16 | |
| $CH_2ClCN$ | | 161.22 | 2.47 |
| | $CCl_4$ | 159.69 | |
| | $C_6H_{12}$ | – | |
| | DMF | 150.40 | |
| $CH_3Cl$ | | 149.64 | 1.82 |
| | $CCl_4$ | 149.18 | |
| | $C_6H_{12}$ | 148.58 | |
| | DMF | 152.14 | |
| $CH_3Br$ | | 151.44 | 1.60 |
| | $CCl_4$ | 150.98 | |
| | $C_6H_{12}$ | 150.54 | |
| | DMF | 151.59 | |
| $CH_3I$ | | 151.09 | 1.28 |
| | $CCl_4$ | 150.65 | |
| | $C_6H_{12}$ | 150.31 | |
| | DMF | 135.99 | |
| $CH_3CN$ | | 136.15 | .33 |
| | $CCl_4$ | 135.66 | |
| | $C_6H_{12}$ | – | |
| | DMF | 133.77 | |
| $CH_3CCl_3$ | | 133.46 | .52 |
| | $CCl_4$ | 133.31 | |
| | $C_6H_{12}$ | 133.25 | |

Their study of bromoform was extended to include thirty organic solvents with a wide variety of function groups. For the aliphatic solvents $\Delta J$ from cyclohexane to the solvent in question gave the following order of magnitudes:

$$\text{Halogen} < \text{CHO} = NR_2 < OH < OR < C{=}O = NHR < NH_2$$

It is noteworthy that the quite polar compounds $CH_3CN$ and $CH_3CCl_3$ which are not expected to hydrogen bond, but which do have significant dipoles and hence might display reaction field effects show $\Delta J$ values of 0.33 and 0.52 Hz respectively; precisely the range of values predicted by Raynes [15].

Table 4. *Solvent effects on* $^1J_{C-H}$ *in chloromethyl ethers*

| Solvent | $^1J_{CH}$(Hz) | |
|---|---|---|
| | $ClCH_2OCHMe_2$ | $ClCH_2OCH_2CH_2Cl$ |
| Cyclohexane | 173.0 | 173.5 |
| Neat | 175.0 | 176.0 |
| Tetrahydrofuran | 175.5 | 176.5 |
| Acetone | 176.0 | 177.5 |
| Dimethylformamide | 177.0 | 178.0 |
| Hexamethylphosphortriamide | 177.0 | 178.0 |

Martin, Castro and Martin [22] found increases in $^1J_{C-H}$ for two chloromethyl ethers (Table 4) which are in agreement with the Watts and Goldstein data. Cox and Smith [23] examined $^1J_{^{13}C-H}$, $^1J_{^{13}C-^{19}F}$ and $^2J_{H-F}$ in difluoro- and trifluoromethane (Table 5). The results for the $^{13}$C-H coupling are in accord with those

Table 5. *Solvent dependence of coupling constants in difluoro- and trifluoromethane*[a]

| Solvent | $J_{^{13}C-H}$ | $J_{^{13}C-F}$ | $J_{HF}$ | Conc. mole % |
|---|---|---|---|---|
| $CH_2F_2$ | $(184.5)$[b] | $(234.8)$[b] | $(50.22)$[b] | |
| Cyclohexane | 181.60 | – | 50.22 | 5 |
| DMSO | 187.15 | – | 50.10 | 5 |
| Cyclohexane | 182.10 | –236.58 | 50.1 | 10 |
| Acetone | 184.10 | –232.78 | 50.3 | 10 |
| DMSO | 186.50 | –232.12 | 50.2 | 10 |
| $CHF_3$ | $(239.1)$[b] | $(274.3)$[b] | $(79.23)$[b] | |
| $CCl_4$ | 238.10 | –274.22 | 79.25 | 15 |
| Acetone | 245.35 | –274.12 | 79.25 | 15 |
| DMSO | 247.30 | –275.22 | 79.30 | 15 |

[a] Probable errors are 0.2 Hz or less for all values reported here except the $^{13}$C-F couplings for which the probable errors are 0.4 Hz.
[b] Values reported by Frankiss, Ref. 63, for 95 % solutions in cyclohexane.

of Watts and Goldstein although the $\Delta J$ values are a bit smaller than might be expected by comparison with electronegativities and $\Delta J$ values for the other halomethanes. Surprisingly, the $^{13}C$-$^{19}F$ coupling constant increases by 4.46 Hz in difluoromethane, but remains constant for two solvents and decreases by 1.0 Hz for the third in trifluormethane. The difluoromethane result is consistent with other work (*vide infra*, Ref. 14), but the trifluoromethane result is not. Most surprisingly, the two-bond H-F coupling constant is solvent invariant in both compounds! This result is totally unexpected!

These results may be qualitatively rationalized using Eq. (1). Assume some electric field, either a reaction field or from hydrogen bonding, which is oriented in such a way as to favor a shift of electrons from hydrogen towards fluorine. A shift of electrons away from hydrogen towards carbon produces an increase in the contribution of the carbon $s$-orbitals to the C-H bond and a corresponding increase in $J_{13_{C-H}}$. The larger change for $J_{13_{C-H}}$ in trifluoromethane is qualitatively explained since we expect a larger shift of electrons away from hydrogen in that compound. The invariance of $J_{HF}$ occurs because a decrease in the contribution of a particular atomic orbital to the MO at atom A (hydrogen) ($C_{iA}$ and $C_{jA}$ decrease) is exactly balanced by an increase in the contribution of the atomic orbital to the MO at atom B (fluorine) ($C_{iB}$ and $C_{jB}$ increase) so that the product of these terms for any pair of orbitals divided by the energy difference between the orbitals is constant. Obviously, the energies of the orbitals will also change. This change is expected to be small and is probably less important than the changes in the coefficients. This qualitative explanation is supported by the observed solvent dependence of $J_{13_{C-F}}$. In difluoromethane the shift of charge in more polar solvents acts to decrease the contribution of fluorine $p$-orbitals to the C-F bond, effectively increasing the $s$-contribution to that bond and producing an algebraic increase in $J_{13_{C-F}}$. The redistribution process effectively moves charge from two hydrogens to two fluorines. In trifluoromethane the larger increase in the $s$-contribution to the single C-H bond (compared with $CH_2F_2$) necessarily means that a decrease is expected in the $s$-contribution to the three C-F bonds (again relative to the change in $CH_2F_2$). In this case the decreased $p$-contribution of fluorine is not sufficient to offset the decrease in $s$-contribution resulting from the increase in the $s$-contribution to the C-H bond. Thus, $\Delta J_{13_{C-F}}$ in $CHF_3$ is expected to be smaller than in $CH_2F_2$, exactly the result which is observed. The solvent invariance of $^2J_{H-F}$ is again rationalized by assuming that the product of the coefficients for H and F remains constant.

De Jeu, Angad Gaur and Smidt[24] suggest an indirect hydrogen bonding effect to rationalize the changes found for $^1J_{C-H}$ in acetone and dimethylsulfoxide (Table 6). This interpretation is based on the linear relationship found between $^1J_{C-H}$ and the $^{17}O$ chemical shift in the same solvent. Hydrogen bonding between solvent and the carbonyl oxygen increases the importance of dipolar resonance structures for the carbonyl group. The increased positive charge on

Table 6. *Solvent effects on* $^1J_{C-H}$ *in acetone and dimethylsulfoxide*

| Solvent | $^1J_{CH}(\pm 0.2$ c/s) | |
| | Me$_2$CO | Me$_2$SO |
|---|---|---|
| Carbon tetrachloride | 126.7 | 137.6 |
| Carbon disulfide | 126.6 | – |
| Acetone | 126.6 | – |
| Nitrobenzene | 127.0 | – |
| Acetonitrile | 127.1 | – |
| Aniline | 127.3 | – |
| Chloroform | 127.4 | – |
| Water | 127.4 | 139.2 |
| Phenol | 128.0 | 138.0 |
| Acetic acid | 128.1 | – |
| Water | 128.1 | – |

carbon causes an increase in $^1J_{C-H}$ by mechanisms similar to those proposed to explain the effect of electronegative groups in increasing $^1J_{C-H}$. Polar solvents interact with the carbonyl group via dipole-dipole interactions producing effectively the same result.

All the above studies have dealt with $^{13}$C-H couplings in $sp^3$ hybridized systems. A few investigators have studied the solvent dependence of $^{13}$C-H couplings in $sp^2$ hybridized systems. Such systems offer the advantages of being rigid and possibly less prone to specific interactions. Watts *et al.* [25] examined $^1J_{^{13}C-H}$, of *cis* and *trans* 1,2-dichloroethylene, *cis* and *trans* 1,2-dibromoethylene and 1,1-dichloroethylene in cyclohexane and DMF (Table 7). Several qualitative

Table 7. *Comparison of NMR*[a] *and IR*[b] *solvent dependence for* $J_{^{13}C-H}$ *of disubstituted ethylenes*

| Compound | $\Delta J$(C-H) | $\Delta\bar{\nu}$C-H | $\Delta\bar{\nu}$C-X |
|---|---|---|---|
| C$_2$HCl$_3$ | 3.40 | – | – |
| *cis*-C$_2$H$_2$Cl$_2$ | 2.97 | 9.6 | 9.7 |
| *trans*-C$_2$H$_2$Cl$_2$ | 2.40 | 16.8 | 10.5 |
| *cis*-C$_2$H$_2$Br$_2$ | 2.70 | 19.9 | 6.2 |
| trans-C$_2$H$_2$Br$_2$ | 2.00 | 19.6 | 8.5 |
| 1,1-C$_2$H$_2$Cl$_2$ | 0.73 | 14.2 | 3.5 |

a) Values in cps; concentrations are 50 mole per cent.
b) Values in cm$^{-1}$; concentrations are 4–9 mole per cent.

trends are evident. The more halogens the greater the effect on $J_{13_{C-H}}$. For di-halocompounds the order of $\Delta J$ is *cis*>*trans*>1,1-. This suggests a specific effect localized at the $\alpha$ halogen. Support for this hypothesis is found in the rough correlation between the solvent effect on the C-X stretching frequency and $\Delta J$.

Investigation of the concentration dependence of $J_{13_{C-H}}$ of trichloroethylene and *cis* 1,2-dichloroethylene (Table 8) over the range of 50 – 100 mole % showed a small but significant change. Both *cis* and *trans* isomers of the dichloro- and dibromoethylenes show the same slopes for the concentration dependence (Figs. 1 and 2). This is inconsistent with the idea of a reaction field effect since

Table 8. *Concentration dependence of* $^1J_{13_{C-H}}$ *for haloethylenes in cyclohexane or dimethylformamide*

| Solvent | Concentration of solute (mole %) | $^1J_{CH(\pm0.2\,c/s)}$ |
|---|---|---|
| | cis-$C_2H_2Cl_2$ | |
| Neat | 100 | 197.89 |
| $C_6H_{12}$ | 91.9 | 197.72 |
| $C_6H_{12}$ | 81.9 | 197.50 |
| $C_6H_{12}$ | 77.1 | 197.45 |
| $C_6H_{12}$ | 68.9 | 197.35 |
| $C_6H_{12}$ | 53.5 | 197.17 |
| $C_6H_{12}$ | 49.0 | 197.12 |
| $Me_2NCHO$ | 79.3 | 198.64 |
| $Me_2NCHO$ | 71.3 | 199.27 |
| $Me_2NCHO$ | 59.1 | 199.86 |
| $Me_2NCHO$ | 43.8 | 200.70 |
| $Me_2NCHO$ | 41.8 | 200.62 |
| | Trichloroethylene | |
| Neat | 100 | 200.92 |
| $C_6H_{12}$ | 85.6 | 200.74 |
| $C_6H_{12}$ | 70.3 | 200.68 |
| $C_6H_{12}$ | 67.1 | 200.73 |
| $C_6H_{12}$ | 49.2 | 200.50 |
| $Me_2NCHO$ | 81.2 | 201.88 |
| $Me_2NCHO$ | 74.7 | 202.74 |
| $Me_2NCHO$ | 68.8 | 202.98 |
| $Me_2NCHO$ | 49.7 | 204.20 |
| $Me_2NCHO$ | 37.2 | 204.48 |

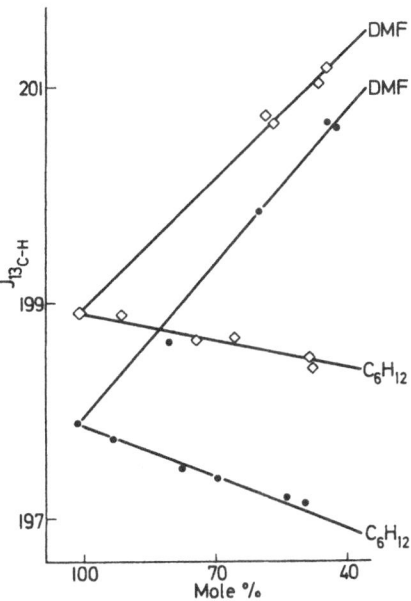

Fig. 1. Concentration dependence of $^1J_{13_{C-H}}$ for dichlorethylenes in cyclohexane and dimethylformamide; ● = *cis-*; ♦ = *trans-*

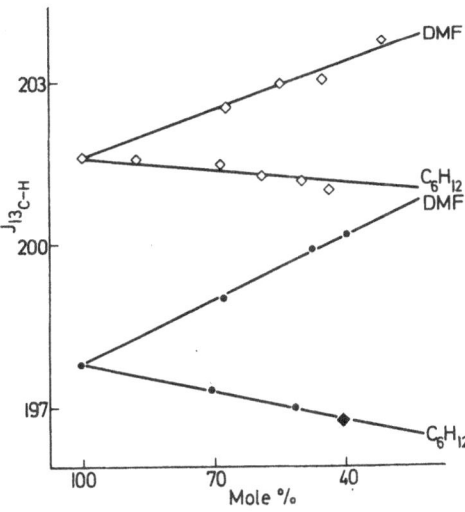

Fig. 2. Concentration dependence of $^1J_{13_{C-H}}$ for dibromethylenes in cyclohexane and dimethylformamide; ● = *cis-*; ♦ = *trans-*

137

the *cis* isomers (which have dipole moments) should show appreciably larger concentration dependence than the *trans* isomers. In fact, if the reaction field were the only way to obtain a solvent effect on $J_{13_{C-H}}$ the trans *isomers* would be expected to show very little if any solvent effects.

It is interesting that the *difference* between $\Delta J$ for the *cis* isomer and $\Delta J$ for the *trans* isomer (Table 7) is approximately 0.7 Hz in each case; about the same as the value of $\Delta J$ observed for the 1,1-isomer. For either *cis-trans* pair it might be assumed that both isomers experience a solvent- C-X bond interaction, whereas the *cis* isomer also experiences a reaction field effect. The 1,1 isomer also experiences C-X interaction with the solvent, but since the halogens are removed from the vicinity of the C-H bonds this has little effect on $^1J_{13_{C-H}}$. Thus, $\Delta J$ for the 1,1-isomer reflects only a reaction field effect. Further, the 0.7 Hz change or difference is the same range as the theoretically predicted effect of 0.3 – 0.5 Hz for a "through space" electric field.

Bell and Danyluk [14,26)] reported the solvent dependence of both $^1J_{13_{C-H}}$ and $^1J_{13_{C-19_F}}$ in *cis*- and *trans*-1,2-dichlorofluoroethylene (Table 9). The change in $^1J_{13_{C-H}}$ (6.9 Hz) is the same for both isomers and is roughly twice the value found by Watts and Goldstein (Table 7) for trichloroethylene. This is consistent with the trend towards larger $\Delta J$ values with increased halogen substitution. However, differences in the solvents used and in concentrations do not permit such data to be used to obtain a quantitative measure of the substituent effect. The changes in $^1J_{13_{C-F}}$ are about half the magnitude of the changes in $^1J_{13_{C-H}}$ and, significantly, are in the opposite direction. The $^{13}$C-H coupling displays an apparent *increase* while the $^{13}$C-$^{19}$F coupling shows an apparent *decrease* in a series of solvents of increasing polarity. This information provided some of the first real evidence that solvent effects on couplings might be related to the absolute sign of the coupling constant and might be rationalized on the basis of theoretical models for the coupling phenomenon.

The reaction field component along the C-H and C-F bonds ought to have the same direction. Using the contact mechanism as formulated in Eqs. (1) and (7), the field in the region of the C-H will "tend to increase the *s* character of the C-hybrid bonding orbital" and hence will lead to a more positive value for $^1J_{13_{C-H}}$. Since $^1J_{13_{C-H}}$ is positive the experimental magnitude of the coupling will increase. The reaction field acting in the region of the C-F bond will shift the bonding electrons closer to the fluorine. This increased polarity causes the $^{13}$C-$^{19}$F coupling to become more positive. Since the $^{13}$C-$^{19}$F coupling is negative a positive shift results in a decrease in the experimentally observed magnitude of the coupling constant. While the theoretical explanation is qualitative the implication is clear. An increased electric field (presumably appropriately oriented) is expected to produce a positive shift in the coupling constant. Thus,

Table 9. *Solvent dependence of coupling constants in cis- and trans-1,2-dichlorofluoro-ethylene*

| Solvent | $^1J_{CH}$ | $^1J_{CF}$ | $^2J_{CCH}$ | $^2J_{CCF}$ | $^3J_{HF}$ |
|---------|-----------|-----------|------------|------------|-----------|
| *cis* isomer | | | | | |
| Cyclohexane | 195.6 | 300.0 | – | 53.6 | 3.80 |
| 1,4-Dioxane | – | – | – | – | 4.23 |
| Benzene | 198.0 | – | 5.1 | – | 4.10 |
| Carbon disulfide | 195.5 | 300.1 | 5.3 | 53.0 | 4.00 |
| Isopropyl ether | 199.6 | – | 5.1 | – | 4.00 |
| Chloroform | 199.6 | 300.0 | 6.0 | 53.0 | 4.05 |
| Acetone | 199.5 | – | 5.6 | – | 4.40 |
| Methanol | – | – | – | – | 4.10 |
| Acetonitrile | 198.0 | 298.6 | 7.5 | 53.8 | 4.30 |
| Dimethylsulfoxide | 201.0 | 298.3 | – | 53.7 | 4.30 |
| N, N-Dimethylformamide | 202.5 | 297.7 | – | 53.1 | 4.60 |
| *trans* isomer | | | | | |
| Cyclohexane | 201.7 | 307.0 | – | 20.0 | 17.45 |
| 1,4-Dioxane | – | – | – | – | 19.30 |
| Benzene | 204.0 | – | 11.0 | – | 18.55 |
| Carbon disulfide | 201.8 | 306.6 | 11.0 | 20.0 | 18.00 |
| Isopropyl ether | 205.2 | – | 11.1 | – | 18.45 |
| Chloroform | 203.8 | 306.2 | 11.1 | 20.0 | 18.10 |
| Acetone | 205.9 | – | 11.1 | – | 19.60 |
| Methanol | – | – | – | – | 18.80 |
| Acetonitrile | 204.5 | 304.5 | 11.1 | 19.2 | 19.00 |
| Dimethylsulfoxide | 207.1 | 304.6 | – | 18.0 | 20.40 |
| N, N-Dimethylformamide | 208.6 | 303.4 | 11.7 | 17.9 | 20.20 |

experimental observation of an apparent increase or decrease in the one-bond coupling in a series of solvents of increasing polarity indicates that the observed coupling constant is absolutely positive or negative respectively.

Few other studies are available on one-bond $^{13}$C-H or $^{13}$C-$^{19}$F coupling constants as a function of solvent. Rahkamaa and Jokisaar [27] reported the solvent and temperature dependence of $^1J_{13_{C-H}}$ in ethyl formate in three solvents; CCl$_4$, CS$_2$ and acetone. The increases of 1.3–2.0 Hz depending on the temperature were about half the size of the temperature effects. Both were attributed to self association of the ethyl formate. Dhingra *et al.* [28] reported that $^1J_{13_{C-F}}$ of trifluoroacetic acid increases with decreasing concentration in CCl$_4$, dioxane, acetone, water and acetonitrile. The change was apparently in order of the dipole moments of the solvents being largest ($\sim 6$ Hz) for acetonitrile. In both cases, the limited data available might be solvent effects or might result from ionization, in the case of trifluoroacetic acid, conformational changes for the ester, etc.

In summary, $^{13}$C-H couplings always increase in solvents of increasing polarity (as measured by basicity, dipole moment or dielectric constant). The magnitude of the change is proportional to the number and kind of electronegative substituents and varies between 2 and 5 % of the value for the pure compound. The principle interaction mechanism seems to be specific interactions, primarily hydrogen bonding, but there is a residual reaction field effect on the order of 0.5–1 %. The fact that there is an apparent correlation with the electronegativity of the substituents might also be a result of increased dispersion interactions resulting from the fact that the more heavily substituted systems also have considerably more electrons than the less heavily substituted systems. Such an effect, if it exists, is obscured by the strong hydrogen bonding effects present in most of the systems studied to date.

With fewer results available it is difficult to make generalizations about $^{13}$C-$^{19}$F couplings. In simple systems they seem to increase algebraically as do the $^{13}$C-H couplings. The net result being an apparent decrease in the observed values. It is not clear whether the contrary results are true solvent effects as considered here or the result of conformational changes and the like.

### 2. $^{29}Si-^{19}F$ Couplings

One of the most significant studies of solvent dependent coupling constants to date is the report of Coyle et al. [29] on the solvent dependence of the $^{29}$Si-$^{19}$F coupling constant in silicon tetrafluoride. This study provides some of the first unambiguous evidence for the effect of dispersion interactions on coupling constants. Silicon tetrafluoride is a centrosymmetric, effectively sperical molecule. It cannot hydrogen bond readily. Neither does it have any dipole moment and hence cannot induce reaction fields. Only dispersion interactions and perhaps *weak* specific interactions are possible.

The data in Table 10 can be compared with values for the coupling in pure SiF$_4$ as a gas (169.00 ± 0.8, 30 Atm; 198.84 ± 0.8 Hz, 110 Atm) and as a liquid (-52 °C, 169.97 ± 0.08 Hz). The coupling increases in all solvents. For a series of related solvents of the formula SiX$_n$F$_{4-n}$ or CX$_n$F$_{4-n}$ the Si-F coupling constant of the solute increases monotonically with $n$. Specific intermolecular fluorine-fluorine interactions were suggested to explain this phenomenon.

Subsequently, Hutton, Bock and Schaefer [29], using Coyle's data, showed that $^1J_{^{29}Si-^{19}F}$ correlated very well with dispersion forces as indicated by a plot of the coupling constant vs the heat of vaporization at the boiling point for the solvents used. Unfortunately, the intercept for their data ($H_b = 0$) is $\sim 164$ Hz which is lower than the gas phase value of 169 Hz. Laszlo and

Table 10. $J_{Si-F}$ in $SiF_4$ (15 mole % or less)

| Solvent | $J$ | Solvent | $J$ |
|---------|-----|---------|-----|
| $Si_2OF_6$ | 170.51 | $SiF_2Br_2$ | 174.51 |
| $CF_3CN$ | 170.66 | $(CH_3)_4Si$ | 174.68 |
| $CClF_3$ | 170.78 | $CCl_3F$ | 175.03 |
| $CH_3SiF_3$ | 171.12 | $CH_2Cl_2$ | 175.23 |
| $SiF_3Br$ | 171.51 | $CHCl_3$ | 176.12 |
| $C_2H_5SiF_3$ | 172.01 | $Si_2Cl_5F$-$Si_2Cl_6$(1:3) | 176.14 |
| $CH_2$=$CHSiF_3$ | 172.05 | $SiFBr_3$ | 176.45 |
| $(CH_3)_2SiF_2$ | 172.35 | $CCl_4$ | 176.83 |
| $(CH_3)_3SiF$ | 173.06 | Cyclo-$C_6H_{12}$ | 176.88 |
| $C_6F_6$ | 173.44 | $C_6H_6$ | 176.98 |
| $CCl_2F_2$ | 173.67 | $BBr_3$ | 178.0 |
| $(C_2H_5)_2O$ | 173.70 | $SiBr_4$ | 178.61 |

Speert [31] also found a good correlation with the McRae dispersion term (for those solvents whose refractive index was known) with an intercept ($n = 0$) of $161.15 \pm 2$ Hz; again quite low in comparison with the gas phase value.

It is unfortunate that Coyle *et al.* used the relatively esoteric solvents listed in Table 10. The combination of a one-bond coupling, gas phase values for the coupling, and an extensive list of solvents of known properties would provide a wonderful opportunity to evaluate the Raynes partition approach (Eq. 8). Even so, Raynes [12] was able to develop a relationship between $\Delta J_{29_{Si}-19_F}$ and

Table 11. *Observed and calculated $J_{Si-F}$ values for $SiF_4$ on going from the gas phase to solution in a series of solvents*

| | $\Delta J$(obs.) | $\Delta J$(calc.) | Solvent | $\Delta J$(obs.) | $\Delta J$(calc.) |
|---|---|---|---|---|---|
| $CH_2Cl_2$ | 6.31 | 6.42 | $SiF_4$ | 1.05(−52 °C) | 1.00 |
| $CHCl_3$ | 7.20 | 7.17 | $SiF_3Br$ | 2.59 | 3.17 |
| $CCl_4$ | 7.91 | 7.92 | $SiF_2Br_2$ | 5.59 | 5.34 |
| | | | $SiFBr_3$ | 7.53 | 7.51 |
| $CF_3Cl$ | 2.48(−3 °C) | 2.52 | $SiBr_4$ | 9.69 | 9.68 |
| $CF_2Cl_2$ | 4.75 | 4.32 | | | |
| $CFCl_3$ | 6.11 | 6.12 | $SiF_3Me$ | 2.20 | 2.19 |
| | | | $SiF_2Me_2$ | 3.43 | 3.38 |
| $Si(Cl,F)1:3$ | 2.97 | 2.91 | $SiFMe_3$ | 4.14 | 4.57 |
| $Si(Cl,F)1:1$ | 4.81 | 4.82 | $SiMe_4$ | 5.76 | 5.76 |

Table 12. *Empirical bond contribution parameters for solvent dispersion interaction effects on* $^1J_{29_{Si}-19_F}$ *from Raynes*

| | |
|---|---|
| $j(C-F) = 0.18$ | $j(Si-F) = 0.25$ |
| $j(C-H) = 1.23$ | $j(Si-Me) = 1.44$ |
| $j(C-Cl) = 1.98$ | $j(Si-Cl) = 2.16$ |
| | $j(Si-Br) = 2.42$ |

a series of empirical parameters of the form $J(C-X)$ and $J(Si-X)$ for the solvents listed in Table 11. The result is quite good, although one is reminded of the old statistician's story that three or four adjustable parameters will suffice to describe an elephant and a few·more will permit a description of the Universe. The parameters (Table 12) assume somewhat more validity when it is-noted that the values bear a clear relation to the atom polarizabilities as might be expected for dispersion forces. The qualitative correlation between the rather small value for $J(C-F)$ compared to $J(C-H)$ nicely explains the different $\Delta J$ values in $C_6F_6$ and $C_6H_6$ (Table 10) of 4.52 Hz and 8.06 Hz respectively.

### 3. $^1J_{31_{P}-H}$ and $^1J_{31_{P}-19_F}$

Considering the importance of $^{31}P$ NMR spectroscopy it is surprising that very few examples of solvent dependent $^{31}P-X$ couplings are known. Ebsworth and Sheldrick [32)] report the solvent and temperature dependence of $^1J_{31_{P}-H}$ of phosphine in eight solvents (Table 13). Hydrogen bonding seems the obvious causative factor for solvents such as ammonia and chloroform. The remaining solvents show increases in $J$ roughly in order of increasing dielectric constant, although a correlation with increases in refractive index is about equally likely. The increase in $^1J_{P-H}$ (which is positive) with increased dielectric constant, increased refractive index, decreased temperature, or with hydrogen bonding are all explained internally as a variation in the relative contribution of the phosphorus $3s$ orbital to the lone pair and to the bonding orbitals, in accord with the model presented in Sect. II and elaborated for carbon-hydrogen couplings.

Fields *et al.* [33)] examined the closely related bis (trifluoromethyl) phosphine (Table 14) and found a similar increase in $^1J_{P-H}$ with increasing polarity of the solvent. They noted a correlation between $J_{P-H}$ and the proton chemical shift (confidence limit of the correlation coefficient was 99.9 %). Again hydrogen bonding was suggested as the principle causative factor since correlations with dielectric constant or refractive index were not found. The two-bond $^2J_{P-F}$ was noted to decrease while the three-bond $^3J_{H-F}$ coupling constant was solvent invariant (*vide infra*).

Table 13. *Solvent and temperature dependence of $^1J_{P-H}$ in phosphine*

| Solvent | T °C | J c/sec |
|---------|------|---------|
| Benzene | +21 | 186.6±0.2 |
|         | +4  | 187.1±0.2 |
| Trimethylamine | +21 | 183.9±0.3 |
|         | 0   | 184.3±0.2 |
|         | -24 | 184.8±0.2 |
| Acetonitrile | +21 | 189.0±0.2 |
|         | 0   | 189.4±0.2 |
|         | -24 | 190.1±0.2 |
| Carbon tetrachloride | +21 | 184.9±0.2 |
|         | 0   | 185.4±0.2 |
|         | -24 | 186.2±0.2 |
| Cyclopentane | +21 | 183.0±0.3 |
|         | +1  | 183.3±0.4 |
|         | -22 | 183.4±0.2 |
| Ammonia | +21 | 189.9±0.2 |
|         | 0   | 191.2±0.2 |
|         | -24 | 192.3±0.2 |
| Chloroform | +21 | 189.2±0.2 |
|         | +2  | 190.0±0.1 |
|         | -23 | 191.0±0.1 |
| Carbon disulfide | +21 | 185.6±0.2 |
|         | +3  | 185.8±0.2 |
|         | -24 | 186.5±0.2 |

Table 14. *Solvent dependence of NMR parameters in bis(trifluoromethyl)phosphine*

| Solvent | $\delta(^1H)$ (p.p.m.) | $^1J_{PH}$ (Hz) | $^2J_{PF}$ (Hz) | $^3J_{HF}$ (Hz) |
|---------|------------------------|-----------------|-----------------|-----------------|
|         | (±0.02) | (±0.8) | (±0.3) | (±0.2) |
| Nil | 4.57 | 217.0 | 70.0 | 9.9 |
| Tetramethylsilane | 4.65 | 216.1 | 69.4 | 9.7 |
| Carbon tetrachloride | 4.71 | 217.2 | 68.1 | 9.7 |
| Chloroform | 4.73 | 220.0 | 67.4 | 9.8 |
| Fluorotrichloromethane | 4.64 | 216.0 | 69.2 | 9.7 |
| Acetonitrile | 5.09 | 236.0 | 64.7 | 10.1 |
| Benzene | 3.62 | 225.2 | 66.1 | 9.9 |
| Acetone | 5.33 | 239.0 | 63.9 | 10.3 |
| Ether | 5.01 | 229.1 | 67.3 | 10.0 |

Kleiman, Morkovin and Ionin [34] report changes of 2-3 Hz for $^1J_{\text{P-H}}$ of dimethyl hydrogen phosphite in polar and nonpolar organic solvents and extreme values of 743.8 Hz in trifluoroacetic acid and 627.9 Hz in ammonia buffer at pH 11.4. A temperature effect (1 : 1 solution in toluene) of 8.7 Hz (-80° to +120 °C) was also found. It seems likely that these results illustrate *strong* hydrogen bonding or perhaps dimer-monomer equilibria involving the P=O group which in turn drastically affects the s-character of the orbitals involved in the P-H bond. Similar but more criptic results are given by Vinogradov *et al.* [35].

An increase from 1404 Hz in the pure gas to 1423 Hz in $CCl_4$ is reported by Raynes *et al.* [36] for the $^1J_{\text{P-F}}$ coupling constant in $PF_3$. Dispersion interactions and electric field effects are suggested as causative factors.

### 4. Other One-Bond Couplings, Ions

Paolillo and Becker [37] report the solvent dependence of $^1J_{^{15}\text{N-H}}$ for 0.1 M aniline in several solvents (Table 15). Hydrogen bonding, either of the $NH_2$ protons to the solvent or of solvent protons to the lone pair of the $NH_2$ group, (or both) is probably the major causative factor.

Table 15. *Solvent dependence of* $^1J_{^{15}\text{N-H}}$ *in aniline*

| Solvent | $J_{^{15}\text{N-H}}$ |
| --- | --- |
| Cyclohexane-$d_{12}$ | 78.0 |
| Carbon tetrachloride | 78.0 |
| Deuterochloroform | 78.0 |
| Dioxane-$d_8$ | 80.6 |
| Pyridine-$d_5$ | 81.4 |
| Acetone | 82.1 |
| Dimethylformamide-$d_7$ | 82.3 |
| Dimethylsulfoxide-$d_6$ | 82.3 |

Solvent dependent coupling constants have been reported for a number of complex ions, usually fluorides. These reports frequently appear in the course of studies conducted for some other purpose and hence the solvent dependence of the coupling constant has not always received the detailed consideration or explanation it might merit. For example, Kuhlmann and Grant [38] found the $^1J_{^{11}\text{B}-^{19}\text{F}}$ coupling of the tetrafluoroborate ion to vary from 1 to 5 Hz in various aqueous solutions. This was attributed to ion pair formation whereby the observed coupling constant was the time average of two or more species, the free ion, an ion pair, and possibly others. Haque and Reeves [39] disagreed with this explanation, cited evidence against the existence of an inner shell ion pair and proposed hydrogen bonding as the causative factor, at least for water, acetone-water and dioxane-water solutions of tetrafluoroborates. Preferential solvation of tetrafluoroborate by the protolysis product of DMSO was proposed to explain the changes observed in DMSO-water solutions.

Gillespie et al. [40,41)] managed to bring some degree of order out of chaos. First, it was demonstrated that while $NaBF_4$ (studied by Kuhlmann and Grant) forms ion pairs, $AgBF_4$ and $NH_4BF_4$ do not. Then it was established that in the absence of ion pairing there is a solvent effect on $^1J_{^{11}B-^{19}F}$. The earlier investigators apparently encountered a variety of effects, but did not recognize all of them. Careful examination of $BF_4^-$ in mixed solvents revealed an additional source of difficulty. The boron-fluorine coupling constant *changes sign* as a function of the solvent system! In all of the organic solvent-water mixtures studied the coupling constant was observed to decrease to zero then increase again as the mole fraction of the organic solvent is increased. Smooth curves can be obtained only if it is assumed that the coupling constant changes sign as shown in Fig. 3. Thus, $^1J_{^{11}B-^{19}F}$ changes from somewhere around -1 to

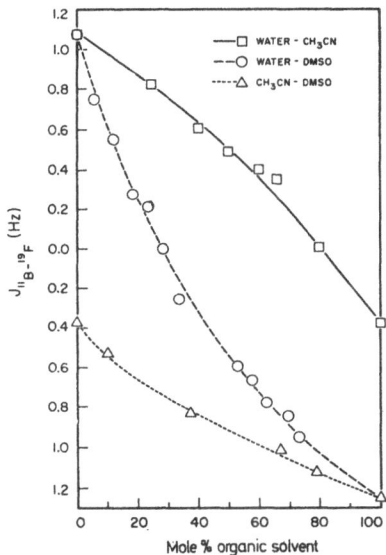

Fig. 3. Solvent dependence of $^1J_{^{11}B-^{19}F}$ for tetrafluoroborate anion in various solvent mixtures

+5 Hz, a total change of 6 Hz, not from 1–5 Hz as originally thought. This is the only case in which it is clearly established that a coupling constant *changes sign* as a function of solvent. Gillespie suggests that the change is caused by some kind of solvent electric field acting according to the model discussed earlier.

Changes of the $^{119}$Sn-$^{19}$F coupling constant in the $SnF_6^{2-}$ ion have been reported by Dean and Evans [42]. Apparently the same kind of solvent-solute interactions are involved. The apparent decrease of $^1J_{119_{Sn}-19_F}$ with increasing polarity of the solvent is consistent with predictions that this coupling is negative.

In all the studies of coupling constant changes in ions the problem seems to be whether the observed changes are the result of ion pairing, specific solvation of the ion, or general bulk solvent effects. The latter two factors cannot be readily distinguished at present. It appears that ion pairing is best investigated by studying the concentration dependence of the coupling constant rather than the solvent dependence.

## 5. Summary of Solvent Effects on One-Bond Couplings

All one-bond coupling constants seem to be solvent dependent. In every case studied so far, with the possible exception of $^1J_{C-F}$ in trifluoromethane and $J_{29_{Si-F}}$ in $SiF_4$, the coupling constants increase in the absolute sense in solvents of greater polarity or higher refractive index. The magnitude of the observed increase is on the order of 1 – 5 % of the value for the coupling constant in the gas phase or in non-interacting solvents, with a few exceptions.

Examples of all types of solute-solvent interactions have been shown with the possible exception of the "magnetic" reaction field. Reaction field effects on one bond couplings, in so far as they can be isolated, seem to be in the range predicted by Raynes (0.3 – 0.5 Hz or a bit more). Specific interactions, particularly hydrogen bonding, appear to have a several fold larger effect. Dispersion interaction effects appear to be about the same magnitude as the reaction field effects. Unfortunately, none of the studies described above were conducted in such a manner as to permit a clear specification of the relative contribution of different interaction mechanisms to the same coupling constant in the same molecules. Hence, the estimates given above are necessarily qualitative.

## C. Two-Bond Coupling Constants

The question of solvent effects on two-bond coupling constants introduces several new dimensions not present in the consideration of one-bond couplings. Changes in bond angles as well as bond length become important. With three atoms involved the possible complications due to hybridization changes, substituent effects, etc., are increased. Since the molecules being studied are likely to be larger and more complex, the nature of solute-solvent interactions becomes potentially more complicated and less readily determinable. Conformational changes in parts of the solute molecule adjacent to the atoms whose coupling is being studied can have an effect on the coupling constant. Finally, it is possible that the same coupling constant, e.g. the H—C—H coupling, can have different absolute signs in different molecules.

The pragmatic result of these increased complications is that considerably more effort has been expended on simply determining whether different types of couplings in different structural systems are or are not solvent dependent. More attention has been focused on what happens to the solute molecule and relatively less attention has been focused on the detailed elucidation of solute-solvent interaction mechanisms.

## 1. $^2J_{H-H}$

Any attempt at understanding the literature data available concerning solvent effects on geminal H—H coupling constants must take cognizance of one fact. Geminal H—H coupling constants may be either positive or negative. Further, it must be recognized that while this has been known for a long time, it is only recently that we have been able to determine experimentally and have had available such data. Both experimental evidence and theoretical understanding have progressed to the point where we can, with considerable confidence, specify the sign of a geminal H—H coupling constant in some particular structural system.

This situation did not exist at the time early studies of solvent dependent geminal H—H coupling constants were performed. As a result there is some confusion for the reader in distinguishing between reported increases or decreases in the experimentally observed magnitude of a geminal H—H coupling constant (which may be either positive or negative) and similar discussions describing increases or decreases in the absolute sense. The latter situation might reflect an increase in the positive contribution to a coupling constant. If the coupling constant is positive the experimentally observed magnitude of the coupling constant increases. If the coupling constant is negative the experimentally observed magnitude of the coupling constant decreases.

Recognizing that conformational changes in a solute molecule result in changes of the coupling constants it is not surprising that virtually all the studies of solvent effects on geminal H—H coupling constants (and most other couplings involving more than one bond) have been conducted with compounds which are for practical purposes rigid. Even so, the phenomenon went unrecognized for many years. Hindsight reveals numerous cases where solvent effects on $^2J_{H-H}$ were reported, for example, Hutton and Schaefer's [43] study of substituted cyclopropanes, or Shapiro et al. [44] investigations of oximes. Attention was first focused on the subject by the reports of Shapiro et al. studying formaldehyde [45] (actually CHDO), and formaldoxime and its methyl ether [46]. Simultaneously, Watts, Reddy and Goldstein [47] reported the solvent dependence of $^2J_{HH}$ in α-chloroacrylonitrile.

147

Shapiro's data for formaldehyde (Table 16) are quite limited, but clearly show a solvent induced decrease for $^2J_{H-D}$ (which is multiplied by the ratio of the gyromagnetic constants of H and D to yield the H—H coupling) going from TMS to acetonitrile.

Table 16. *Solvent dependence of $^2J_{H-D}$ and $^2J_{H-H}$ in $CH_2O$ and $CHDO$[a)]*

| Solvent | $|J_{HD}|$ (Hz) | $|J_{HH}|$ (Hz) |
|---------|------------------|------------------|
| Tetramethylsilane | 6.52±.02 | 42.42 |
| Tetrahydrofuran | 6.26 | 40.70 |
| Acetonitrile | 6.18 | 40.22 |

[a)] All data were obtained on dilute ($< 5\%$) solutions.

It is barely possible (though unlikely in light of subsequent studies) that the proposed change of $\sim 2$ Hz in the geminal H—H coupling in formaldehyde is an artifact resulting from a seven fold multiplication of the much smaller change observed for the H—D coupling. The results shown in Table 17 for formaldoxime and its methyl ether are unambiguous. In both cases the geminal H—H coupling decreases in solvents of increasing polarity.

Table 17. *Solvent dependence of $^4J_{H-H}$ in formaldoxime and formaldoxime methyl ether ($< 5\%$ concentration)*

| Solvent | $\epsilon$ | Formaldoxime | Formaldoxime methyl ether |
|---------|-----------|--------------|---------------------------|
| $C_6H_{12}$ | 1.96 | — | 9.22±.05 Hz |
| $CCl_4$ | 2.20 | — | 8.80 |
| $(n\text{-}C_4H_9)_2O$ | 2.25 | 9.95 | 8.85 |
| $(C_2H_5)_2O$ | 4.00 | 9.82 | 8.58 |
| $CDCl_3$ | 4.61 | 8.58 | 8.32 |
| $CH_2Cl_2$ | 8.46 | 8.35 | 8.28 |
| $(n\text{-}C_3H_7)_2CO$ | 11.6 | 9.52 | 8.46 |
| $n\text{-}C_4H_9OH$ | 15.7 | 9.16 | 8.11 |
| $C_5H_5OH$ | 22.6 | 9.30 | — |
| $CH_3NO_2$ | 34.8 | 8.32 | 7.80 |
| $CH_3CN$ | 34.8 | 8.72 | — |
| $(CH_3)_2SO$ | 46.0 | 9.15 | — |
| $H_2O$ | 72.3 | 8.18 | 6.96 |
|  |  | 7.67 | 7.12 |
| $D_2O$ | 73.6 | 7.63±.05 Hz | 7.12 |

The change for the methyl ether shows a rough correlation with the dielectric constant of the solvent. The oxime itself does not show a particularly good correlation. It is noteworthy that all three compounds show *apparent decreases* in solvents of increased polarity or dielectric constant.

Watts, Reddy and Goldstein found exactly the opposite effect for $\alpha$-chloroacrylonitrile (Table 18). A rough correlation is noted between the *apparent increase* and the reaction field term $(\epsilon-1)/(2\epsilon+2.5)$.

Table 18. *Solvent dependence of $^2J_{H-H}$ in $\alpha$-chloroacrylonitrile*

| Solvent | $\epsilon$ | $J^0$ |
|---|---|---|
| $\alpha$-Chloroacrylonitrile | – | 2.80 |
| Tetramethylsilane | 1.9 | 1.96 |
| Cyclohexane | 1.97 | 1.96 |
| Bromoform | 4.2 | 2.47 |
| d-Chloroform | 4.6 | 2.41 |
| Iodoethane | 7.42 | 2.53 |
| Ethyl bromide | 8.8 | 2.57 |
| 4-Heptanone | 11.7 | 2.95 |
| Acetone | 19.8 | 3.07 |
| Methanol | 30.7 | 2.93 |
| Dimethylformamide | 35 | 3.19 |
| Nitromethane | 35 | 3.03 |
| Dimethylsulfoxide | 46 | 3.24 |

Subsequently, Watts and Goldstein [48] expanded their initial report. For $\alpha$-chloroacrylonitrile ($\alpha$–CAN) $^2J_{H-H}$ was found to vary monotonically with concentration; decreasing upon dilution in solvents whose dielectric constant is less than that of $\alpha$–CAN and increasing in solvents whose dielectric constant is greater than that of $\alpha$–CAN. More limited data showed a similar *apparent increase* for $^2J_{H-H}$ (at infinite dilution) in a series of vinyl halides (Table 19). Since $^2J_{H-H}$ is known to be negative for the vinyl halides the *apparent increase* is an *algebraic decrease* in the absolute sense.

The trend for $\Delta J$ is in accord with the polarizability of the substituent atom, but opposite to the electronegativity trend. McLauchlan *et al.* [49] also studied $\alpha$–CAN as a function of solvent and temperature. Their results, while differing in detail, are in accord with those of Goldstein's group. In a more recent study Goldstein *et al.* [50] reexamined $\alpha$-chloroacrylonitrile in detail in seven different

Table 19. *Solvent dependence of* $^2J_{H\text{-}H}$ *for monosubstituted vinyl compounds* ($CH_2$=CHX)

| X | Solvent | $^2J_{H\text{-}H}$ | $^{\Delta J}C_6H_{12}\text{-DMF}$ |
|------|---------------------|---------|-------|
| CN | $C_6H_{12}$ | 1.20 | |
| | Neat | 0.91 | |
| | DMF | 0.96 | −.24 |
| F | $C_6H_{12}$ | −3.06 | |
| | DMF | −3.39 | −.33 |
| Cl | $C_6H_{12}$ | −1.28 | |
| | Neat | −1.48 | |
| | DMF | −1.67 | −.39 |
| Br | $C_6H_{12}$ | −1.59 | |
| | Neat | −1.80 | |
| | DMF | −2.05 | −.46 |
| I | $C_6H_{12}$ | −0.88 | |
| | Neat | −1.47 | |
| | DMF | −1.52 | −.74 |

solvents and used the resulting data to develop a more sophisticated model of so-lute-solvent interactions. Basically this approach involves inclusion of both reaction field and specific interactions. The end result is to remove most of the deviations observed in plots of $^2J_{H\text{-}H}$ *vs* the reaction field term. This is accomplished by developing a collision complex model to account for specific interaction effects.

Martin and Martin [51] have reported similar small ($\Delta J$ = −0.2 to −0.45 Hz) changes for $^2J_{HH}$ in a series of alkyl substituted vinyl bromides. The same authors [52] have reported a 1.5 Hz decrease with increasing solvent polarity for the negative geminal coupling constant in allenic ketones of the form RCOCH = C=CH$_2$ where R is methyl or ethyl. Reaction field interactions are suggested.

Danyluk's group [53,54] found small (0.2 to 0.9 Hz) algebraic decreases for $^2J_{H\text{-}H}$ in a series of vinylsilanes (Table 20). For these compounds the magnitude of the effect is clearly related to the polarizability of the solute. The geminal protons involved in the coupling showed different solvent effects for their chemical shifts. Also, the correlation between dielectric constant and coupling constants is not very good. Hence, hydrogen bonding or dipole-dipole interactions were proposed as the solute-solvent interaction mechanism.

Table 20. *Solvent dependence of $^2J_{H-H}$ in vinylsilanes (concentration <15 mole %)*

| Solvent | $^2J_{HH}$ (Hz) |
|---|---|
| $Cl_3SiC(Cl)=CH_2$ | |
| Cyclohexane | 1.86 |
| Carbon disulfide | 1.91 |
| Carbon tetrachloride | 1.98 |
| Neat compound | 2.04 |
| Benzene | 2.44 |
| Acetonitrile | 2.53 |
| Tetrahydrofuran | 2.60 |
| Acetone | 2.65 |
| Dioxane | 2.65 |
| Cyclohexane:acetone (2:1) | 2.42 |
| $(CH_3)_3SiC(Cl)=CH_2$ | |
| Cyclohexane | −0.74 |
| Carbon tetrachloride | −0.88 |
| α-CTMS | −1.01 |
| Benzene | −1.24 |
| Diethyl ether | −1.11 |
| Pyridine-$d_5$ | −1.27 |
| Acetone | −1.49 |
| $(C_6H_5)_3SiC(Br)=CH_2$ | |
| Carbon tetrachloride | 1.33 |
| Neat compound (mp 129 °C) | 1.56 |
| Acetone | 1.77 |
| Acetonitrile | 2.08 |
| $(C_6H_5)_3SiC(C_6H_5)=CH_2$ | |
| Carbon tetrachloride | 2.83 |
| Acetone (1 mole %) | 2.79 |
| Acetonitrile (2 mole %) | 2.65 |

Several other cases of solvent dependent geminal H−H coupling constants in olefins have been reported as parts of other studies [61,62,67,79]. Those results are in accord with the trends described above and will be listed along with data on the H−F coupling constants when they are considered.

All of the data discussed to this point deal with changes in $^2J_{H-H}$ across $sp^2$ hybridized carbon. Some couplings show apparent increases; some show

Table 21. *Solvent dependence of coupling constants in styrene oxide, styrene sulfide and 2,2-dichlorocyclopropylbenzene*

| Solvent | $\epsilon$ | $^2J_{HH}$ | $^3J_{H-H}^{trans}$ | $^3J_{H-H}^{cis}$ |
|---|---|---|---|---|
| **Styrene oxide** | | | | |
| Cyclohexane | 1.99 | 6.00 | 2.38 | 3.93 |
| Carbon tetrachloride | 2.20 | 5.85 | 2.40 | 3.94 |
| Benzene | 2.62 | 5.81 | 2.42 | 4.06 |
| Toluene | 2.35 | 5.79 | 2.39 | 3.99 |
| Deuterochloroform | 4.55 | 5.55 | 2.55 | 4.10 |
| Pyridine | 12.3 | 5.69 | 2.42 | 4.11 |
| Acetophenone | 16.99 | 5.67 | 2.41 | 4.07 |
| Acetone | 19.8 | 5.56 | 2.44 | 4.06 |
| o-Nitrotoluene | 25.15 | 5.68 | 2.46 | 4.06 |
| Nitrobenzene | 32.22 | 5.53 | 2.49 | 4.06 |
| Nitromethane | 35.0 | 5.42 | 2.55 | 4.17 |
| Acetonitrile-$d_3$ | 35.1 | 5.40 | 2.56 | 4.16 |
| Dimethylsulfoxide | 46 | 5.31 | 2.43 | 4.13 |
| Neat | – | 5.63 | 2.48 | 4.08 |
| **Styrene sulfide** | | | | |
| Cyclohexane | 1.99 | –1.15 | 5.44 | 6.42 |
| Carbon tetrachloride | 2.20 | –1.18 | 5.36 | 6.49 |
| Benzene | 2.26 | –1.37 | 5.53 | 6.47 |
| Deuterochloroform | 4.55 | –1.52 | 5.55 | 6.62 |
| Pyridine | 12.3 | –1.41 | 5.63 | 6.46 |
| Acetone-$d_6$ | 19.8 | –1.46 | 5.66 | 6.48 |
| Nitrobenzene | 32.22 | –1.47 | 5.57 | 6.54 |
| Nitromethane | 35.0 | –1.61 | 5.76 | 6.55 |
| Acetonitrile-$d_3$ | 35.1 | –1.54 | 5.62 | 6.58 |
| Dimethylsulfoxide-$d_6$ | 46 | –1.55 | 5.82 | 6.57 |
| Neat | – | –1.34 | 5.62 | 6.55 |
| **2.2-Dichlorocyclopropylbenzene** | | | | |
| Carbon tetrachloride | 1.99 | –7.36 | 8.53 | 10.36 |
| Benzene | 2.26 | –7.40 | 8.64 | 10.74 |
| Deuterochloroform | 4.55 | –7.40 | 8.54 | 10.53 |
| Acetone-$d_6$ | 19.8 | –7.54 | 8.35 | 10.88 |
| Nitrobenzene | 32.2 | –7.56 | 8.42 | 10.80 |
| Dimethylsulfoxide-$d_6$ | 46 | –7.70 | 8.74 | 10.64 |
| Neat | – | –7.42 | 8.24 | 10.86 |

apparent decreases in solvents of increasing polarity. With the exception of the vinylsilane studies, all the solute-solvent interactions were assumed to be reaction field interactions or occasionally hydrogen bonding of some unspecified nature. Before proceeding with more complex solutes containing $sp^3$ hybridized $CH_2$ groups it is desirable to resolve the different apparent directions in which the coupling constants change, and the relationship between those changes and the sign of the geminal coupling.

Smith and Cox [13,55,56] pointed out that positive geminal coupling constants which *apparently* decrease and negative geminal coupling constants which *apparently* increase are, in fact, showing exactly the same behavior in the absolute sense; a decrease in the positive contributions to the coupling constant (or alternatively an increase in the negative contribution). The coupling constants become more negative or decrease algebraically. To test this hypothesis they studied styrene oxide, styrene sulfide and 2,2-dichlorocyclopropylbenzene. The structures of styrene oxide and styrene sulfide are sufficiently similar that whatever solvation phenomena occur should be qualitatively the same for the two molecules. While not olefinic, the three-membered ring systems are sufficiently rigid as to preclude conformational changes affecting the coupling constants. Most important, the geminal H−H coupling constant in styrene oxide is positive while the geminal H−H coupling constant in styrene sulfide is negative. The results presented in Table 21 and plotted in Fig. 4 clearly support the hypothesis that the apparent increases or decreases reported are in fact changes in the same absolute direction. The correlation between $^2J_{H-H}$ and the solvent dielectric constant suggests a reaction field interaction in accord with previous observations, although hydrogen bonding or other specific interactions may play some part as evidenced by the anomalously low values observed for both compounds in $d$-chloroform. Assuming some sort of electric field with

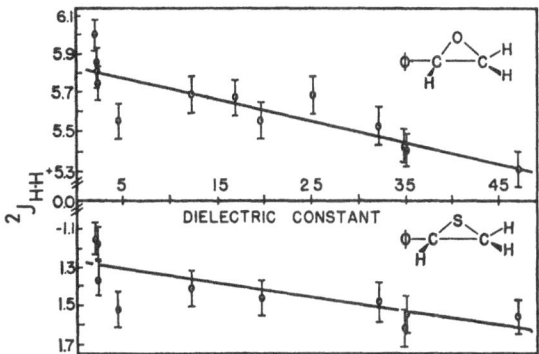

Fig. 4. Plot of $^2J_{H-H}$ for styrene oxide and styrene sulfide versus dielectric constant of the solvent. Vertical bars represent 0.1 Hz error limits

153

a specific orientation with respect to the $CH_2$ group, Smith and Cox [13] formulated a theoretical explanation for the observed decrease of $^2J_{H-H}$ in solvents of increasing polarity.

The simplest explanation is provided by assuming the contact term is the only contributor to the geminal H–H coupling constant and utilizing the theoretical expression (Eq. (1)) developed by Pople and Bothner-By [5]. In the simplest approach the electronic structure of the isolated $CH_2$ group is decribed in terms of four molecular orbitals which are delocalized over both bonds. Two of these are bonding and occupied. The other two are antibonding and unoccupied. The orbitals may be further classified as symmetric or antisymmetric under reflection in the plane perpendicular to the H–C–H plane. Detailed considerations of the effect of substituents on these molecular orbitals led Pople and Bothner-By to propose the following general conclusions for substituent effects on geminal coupling constants:

(1) "Withdrawal of electrons from orbitals *symmetric* between hydrogen atoms (generally inductive effects) should lead to a *positive* change in the coupling constants."

(2) "Withdrawal of electrons from orbitals *antisymmetric* between hydrogen atoms (generally hyperconjugative effects) should lead to a negative change in the coupling constants."

Applications of these rules may be demonstrated using formaldehyde as an example. Inductive withdrawal of electrons by the directly bonded oxygen from the symmetric orbital $\psi_1$ should produce a positive contribution to $^2J_{HH}$ (Rule 1). Concurrent contribution of electrons from the doubly occupied nonbonding $2p_y$ orbital of the oxygen back into the antisymmetric orbital $\psi_2$ should also produce a positive contribution to $^2J_{HH}$ (Rule 2). These shifts are illustrated schematically in Fig. 5. Thus, $^2J_{H-H}$ for formaldehyde is predicted to be large and positive, as is in fact the case [45]. Entirely analogous considerations may be developed for the compounds studied in this paper.

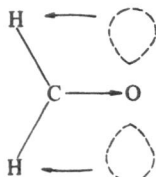

Fig. 5. Schematic illustration of electron transfer in formaldehyde

The qualitative example presented above describes the isolated formaldehyde molecule. In solution the large permanent dipole moment or hydrogen bonding of formaldehyde will induce an appreciable solvent electric field whose orientation in the molecular coordinate system is fixed (presumably parallel to the dipole moment). In the formaldehyde example, the dipole orientation (and hence the electric field orientation) is such as to produce a shift of electrons out of both the symmetric orbital $\psi_1$ and the antisymmetric orbital $\psi_2$. Removal of electrons from $\psi_1$ (pseudo-inductive effect, Rule 1) would cause $^2J_{H-H}$ to become more positive, whereas removal of electrons from $\psi_2$ (pseudohyperconjugative effect, Rule 2) would cause $^2J_{H--H}$ to become more negative. The reasonable assumption that the antisymmetric orbital $\psi_2$ involved in $\pi$-like hyperconjugative interactions is more polarizable than the symmetric orbital $\psi_1$ leads to the conclusion that $^2J_{HH}$ in formaldehyde should become more negative in the absolute sense in the presence of a solvent reaction field. This is exactly the result observed. Similarly, the electric field, perhaps somewhat less favorably oriented, would be expected to decrease the backbonding effect of the nonbonding electrons on oxygen in styrene oxide causing $^2J_{HH}$ to become less positive. In general then, this model predicts that $^2J_{HH}$ should become more negative in solvents of higher dielectric constant.

These generalizations suggest that $\Delta^2J_{HH}$ should be largest for compounds in which a heteroatom having a nonbonding pair of electrons is bound directly to the $CH_2$ group being examined. Formaldehyde, formaldoxime and its methyl ether, which best fit these requirements exhibit large variations of $^2J_{HH}$. Molecules such as styrene oxide having a smaller dipole moment, less favorable dipole orientation, and a smaller hyperconjugative contribution to $^2J_{HH}$ show a smaller variation of $^2J_{HH}$. The proposed model also requires that hydrogen-bonding solvents produce a negative shift in $^2J_{HH}$. Solvents such as chloroform presumably bond to the solute via nonbonding electrons (e.g. at oxygen in styrene oxide), greatly decreasing the hyperconjugative contribution to $^2J_{HH}$ which thus becomes more negative than might be expected for a reaction-field effect alone.

This model predicts that $^2J_{H-H}$ will decrease algebraically for all normal bonding systems, in accord with the data discussed above. Subsequent investigations of other systems have not provided any significant contradictions.

In the same paper Smith and Cox presented some of the first data concerning solvent effects on $^2J_{H-H}$ in $sp^3$ hybridized systems. Two additional variables require consideration when dealing with non-double-bonded systems. Obviously, the hybridization of the carbon involved may be important. Less obvious is the question of dipole moment orientation. The reaction field may be an important factor in affecting the solvent dependence of geminal coupling constants. The reaction field $R$ is usually assumed to be parallel to and roughly proportional to $\mu$, the permanent dipole moment of the solute molecule. In double-bonded compounds hitherto studied, $\mu$, and hence $R$ must lie in the H—C—H plane. In non-olefinic compounds $\mu$ may have a variety of orientations. Conceivably the orientation of the reaction field with respect to the H—C bonds might have a siginificant influence on the magnitude of the observed change in $^2J_{H-H}$. Some distinction between these additional variables is provided by examining the solvent dependence of $^2J_{HH}$ in 1,4-diphenylazetidinone (*1*), styrene carbonate (*2*), and 4-methyl-1,3-dioxolane (*3*):

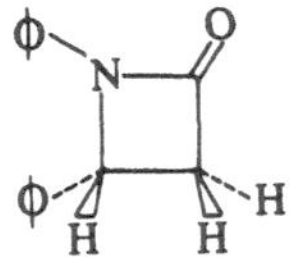

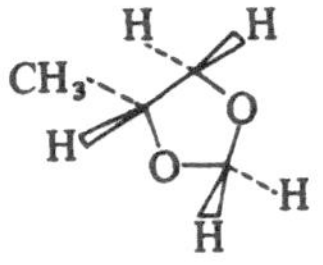

The first two compounds have dipole moments which do not lie in the H—C—H plane. The C-2 methylene in (*3*) is geometrically similar to the methylene in (*2*) and is similarly substituted. However, in the dioxolane the permanent dipole moment must lie in or close to the H—C—H plane. The solvent invariance of $^2J_{H-H}$ in (*1*) and (*2*) in contrast to the solvent dependence of $^2J_{H-H}$ in (*3*) (Table 22) provides some support for this hypothesis.

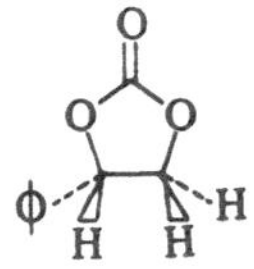

Table 22. *Geminal H—H coupling constant across $C_2$ of 4-methyl-1,3-dioxolane in various solvents*

| Solvent | $\epsilon$ | $^2J_{HH}(C_2)$ |
|---|---|---|
| Cyclohexane | 1.99 | 0.79 |
| Carbon tetrachloride | 2.20 | 0.69 |
| Carbon disulfide | 2.64 | 0.68 |
| Isopropyl ether | 3.88 | 0.73 |
| Deuterochloroform | 4.55 | 0.41 |
| p-Chlorotoluene | 6.08 | 0.61 |
| Methylene chloride | 9.09 | 0.0 |
| Pyridine | 12.3 | 0.0 |
| Acetone | 19.8 | 0.0 |
| Dimethylsulfoxide | 46 | 0.0 |
| Deuterium oxide | 80 | 0.0 |
| N-Methylacetamide | 165 | 0.0 |
| Neat | – | 0.0 |

Smith and Cox [57] extended the study of $sp^3$ hybridized systems to a series of bicyclo-compounds (*4*) with no heteroatom bonded directly to the $CH_2$ group in question.

$$X = Cl, CN, C_6H_5, OAc, OH, CO_2H$$

*4*

In every case $^2J_{H-H}$ decreased by a small amount ($\Delta J$ = -0.53 to -0.13 Hz). The magnitude of the change is consistent with the values predicted by Raynes for an electric field effect, but the limited number of solvents utilized did not permit any useful experimental conclusions concerning solute-solvent interactions.

In the course of a study devoted primarily to the investigation of three-bond H–H coupling constants Fingold [58] reported similar solvent effects on $^2J_{HH}$ in a series of aliphatic heterocyclic compounds (*vide infra*, Table 35). Rattet, Williams and Goldstein [59,60] found solvent dependent geminal H–H coupling constants in open chain systems. Advantage was taken of the chemical shift nonequivalence of the $CH_2$ group protons in diethyl acetals. Again, small decreases ($\Delta J$ = -0.231 to -0.559 Hz) were observed. Larger values were observed for di- and tri-chloroacetal and the smallest value for acetal. Concentration studies on diethyl dichloroacetal revealed essentially the same pattern observed previously by Goldstein's group in their study of α-chloroacrylonitrile. The change in $^2J_{H-H}$ as a function of solvent was monotonic with the reaction field term.

The experimental evidence available concerning solvent effects on geminal H–H coupling constants may be summarized as follows:

(1) Geminal H–H coupling constants always decrease in solvents of increasing polarity (dielectric constant).

(2) The magnitude of the effect is largest ($\sim$ 1 to 2 Hz) for compounds having herteroatoms bonded directly to the $CH_2$ group in question or molecules having large permanent dipole moments with a significant component lying in the H–C–H plane.

(3) For molecules having similar structures and dipole moments the more polarizable molecules exhibit the larger effect.

(4) The principal interaction mechanism *proposed* is reaction field interactions. Specific interaction effects are most noticeable with proton-donor solvents such as chloroform.

The dipole orientation condition is questionable but is supported by other studies on H–F coupling constants. The selection of the reaction field as the primary interaction is also somewhat questionable. Few investigators actually correlated results with the mathematical expressions for the reaction field.

Rather, correlations are presented with the dielectric constant or with "solvent polarity". It is true that the magnitude of the effects observed is frequently in the range of 0.2 – 0.5 Hz as predicted by Raynes. Some evidence for specific interaction is suggested by the collision complex model [50] and by the deviations observed in chloroform solutions, acids and gases. A few investigators have noted that no correlation was found with the refractive index of the solvents suggesting that dispersion forces are not important for $^2J_{H-H}$, at least in those compounds studied.

## 2. $^2J_{H-F}$

Watts and Goldstein [48] reported the solvent and concentration dependence of $^2J_{H-F}$ of vinyl fluoride in cyclohexane and DMF. Smith and Ihrig [61,62] extended those results to other solvents and also measured the solvent dependence of $^2J_{H-F}$ in the isomeric difluoroethylenes and trifluoroethylene (Table 23). The

Table 23. *Solvent dependence of coupling constants in fluoroethylenes*

| Solvent | $^2J_{H-H}$ | $^2J_{H-F}$ | $^3J_{HF}^{trans}$ | $^3J_{HF}^{cis}$ |
|---|---|---|---|---|
| Vinyl fluoride | | | | |
| Cyclohexane | –3.06 | 84.67 | 51.81 | 19.63 |
| Deuterochloroform | –3.34 | 85.49 | 53.61 | 20.53 |
| Acetone | –3.32 | 86.14 | 54.66 | 21.05 |
| Dimethylformamide | –3.39 | 86.47 | 55.45 | 21.56 |
| Trifluoroacetic acid | –3.44 | 86.31 | 54.57 | 21.23 |
| Dimethylsulfoxide | –3.41 | 86.54 | 56.38 | 21.77 |
| | $^2J_{HF}$ | $^3J_{HF}^{trans}$ | $^3J_{FF}^{cis}$ | |
| *cis*-1,2-Difluoroethylene | | | | |
| *n*-Hexane | 71.70 | 19.63 | 18.74 | |
| Cyclohexane | 71.84 | 19.77 | 19.04 | |
| Benzene-$d_6$ | 72.24 | 20.44 | 19.35 | |
| Carbon tetrachloride | 71.80 | 19.99 | 19.02 | |
| Carbon disulfide | 71.76 | 20.04 | 19.66 | |
| Propionic acid | 72.29 | 20.53 | 18.74 | |
| Diethylamine | 72.20 | 20.63 | 19.19 | |
| Diethyl ether | 72.12 | 20.43 | 18.98 | |
| Deuterochloroform | 72.01 | 20.31 | 19.10 | |
| Chlorobenzene | 72.17 | 20.36 | 19.39 | |
| Ethyl acetate | 72.48 | 20.88 | 18.84 | |
| Tetrahydrofuran | 72.42 | 20.84 | 19.24 | |

Table 23 (continued)

| Solvent | $^2J_{H-F}$ | $^3J_{HF}^{trans}$ | $^3J_{HF}^{cis}$ |
|---|---|---|---|
| Methylene chloride | 72.73 | 20.51 | 18.94 |
| Cyclopentanone | 72.56 | 21.06 | 19.35 |
| Acetone-$d_6$ | 72.73 | 21.08 | 18.89 |
| 2-Nitropropane | 72.56 | 20.85 | 18.89 |
| Dimethylformamide | 72.76 | 21.49 | 19.45 |
| Acetonitrile | 72.80 | 21.09 | 18.39 |
| Nitromethane | 72.63 | 21.03 | 18.52 |
| Trifluoroacetic acid | 72.50 | 20.34 | 17.77 |
| Dimethyl sulfoxide | 72.67 | 21.94 | 20.59 |

| trans-1,2-Difluoroethylene | $^2J_{HF}$ | $^3J_{HF}^{cis}$ | $^3J_{FF}^{trans}$ |
|---|---|---|---|
| Cyclohexane | 75.10 | 2.80 | −133.46 |
| Benzene-$d_6$ | 75.14 | 3.04 | −132.72 |
| Carbon tetrachloride | 75.05 | 2.83 | −132.86 |
| Carbon disulfide | 75.06 | 2.84 | −133.79 |
| Propionic acid | 75.17 | 3.15 | −131.57 |
| Diethylamine | 75.20 | 3.17 | −131.88 |
| Diethyl ether | 75.23 | 3.17 | −131.94 |
| Deuterochloroform | 75.14 | 2.96 | −131.96 |
| Chlorobenzene | 75.13 | 2.99 | −132.82 |
| Ethyl acetate | 75.15 | 3.33 | −131.16 |
| Tetrahydrofuran | 75.23 | 3.25 | −131.54 |
| Methylene chloride | 75.17 | 3.08 | −131.54 |
| Cyclopentanone | 75.03 | 3.27 | −131.42 |
| Acetone-$d_6$ | 75.10 | 3.36 | −130.77 |
| Acetonitrile-$d_3$ | 75.30 | 3.33 | −130.20 |
| Trifluoroacetic acid | 75.46 | 3.06 | −130.37 |
| Dimethylsulfoxide-$d_6$ | 75.03 | 3.57 | −130.61 |

| Trifluoroethylene | $^2J_{H-F}$ | $^3J_{H-F}^{cis}$ | $^3J_{H-F}^{trans}$ |
|---|---|---|---|
| Cyclohexane | 70.51 | −4.19 | 12.52 |
| Carbon tetrachloride | 70.59 | −4.18 | 12.57 |
| Carbon disulfide | 70.58 | −4.24 | 12.76 |
| Propionic acid | 70.52 | −4.24 | 13.09 |

Table 23 (continued)

| Solvent | $^2J_{H\text{-}F}$ | $^3J_{HF}^{cis}$ | $^3J_{HF}^{trans}$ |
|---|---|---|---|
| Diethylamine | 70.20 | −4.24 | 13.37 |
| Diethyl ether | 70.28 | −4.27 | 13.09 |
| Deuterochloroform | 70.61 | −4.24 | 12.79 |
| Methylene chloride | 70.73 | −4.27 | 13.00 |
| Cyclopentanone | 70.21 | −4.27 | 13.70 |
| Acetone | 70.39 | −4.31 | 13.53 |
| 2-Nitropropane | 70.43 | −4.30 | 13.26 |
| Dimethylformamide | 70.08 | −4.31 | 14.01 |
| Trifluoracetic acid | 70.92 | −4.26 | 12.63 |
| Dimethyl sulfoxide | 69.93 | −4.30 | 14.32 |
| Propylene carbonate | 70.43 | −4.20 | 13.65 |
| Nitromethane | 70.55 | −4.34 | 13.30 |

most striking feature of these studies is the apparent relationship between the magnitude and sign of the solvent effect on $^2J_{H\text{-}F}$ and the solute dipole moment orientation as shown in Fig. 6. The result is in accord with a reaction field inter-

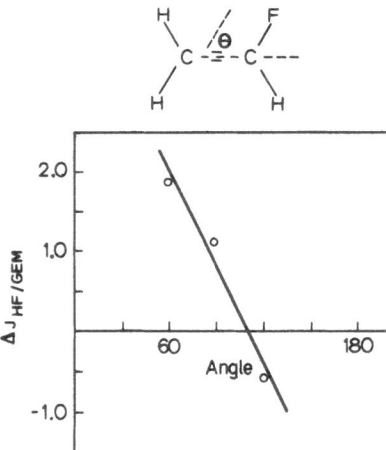

Fig. 6. Plot of $\Delta^2 J_{HF}$ max versus the angle between the solute dipole and a plane bisecting the geminal H−F group

action mechanism. In the case of vinyl fluoride the dipole is oriented so as to produce an electric field which will favor a shift of electrons towards the H and the F of the CHF group resulting in an increase in the positive coupling constant. Trifluoroethylene shows the opposite effect and the *cis* isomer an intermediate change. As expected, the *trans* isomer which does not have a dipole moment shows only small changes which appear to originate from specific interactions. A regression analysis conducted with the *cis*-1,2-difluoroethylene data gives a correlation coefficient between $^2J_{H-F}$ and the reaction field term of 0.89 to 0.93 depending on whether solvents such as benzene and trifluoroacetic acid which might be expected to exhibit specific interactions are excluded from the analysis. This study represents perhaps the best evidence extant for the existence of a reaction field interaction mechanism and the concomitant dipole orientation effect on the solvent dependence of geminal coupling constants. However, the reaction field interaction is clearly not the only contributor to the solvent dependence of $^2J_{H-F}$. In addition to the obvious specific interaction effects noted for some solvents, the regression analysis mentioned above shows that the best fit between $^2J_{H-F}$ and solvent parameters is obtained by including both a reaction field term and a dispersion interaction term (*vide infra*). Also, it should be noted that the vicinal H–F coupling constants, while solvent dependent, do not appear to show any dependence on dipole moment orientation.

The solvent dependence of geminal H–F coupling constants in $sp^3$ hybridized systems has been noted several times, but has received very little detailed study. Evans [18] reported the solvent dependence of $^2J_{H-F}$ in bromochlorofluoromethane (Table 24) which shows a decrease of 1.5 Hz between cyclo-

Table 24. *Solvent dependence of $^2J_{H-F}$ of bromochlorofluoromethane in various solvents (mole fraction CHBrClF = 0.05)*

| Solvent | $^2J_{HF}$ |
|---|---|
| Cyclohexane | $52.1_0$ |
| Nitromethane | $51.5_6$ |
| Acetonitrile | $51.4_3$ |
| Acetone | $51.1_8$ |
| Ether | $51.1_2$ |
| Triethylamine | $50.8_3$ |
| Dimethylsulfoxide | $50.6_7$ |

hexane and DMSO. Specific interaction via hydrogen bonding was suggested as the principal interaction mechanism in accord with the solvent dependence of $^1J_{C-H}$ reported in the same paper. Frankiss [63] reported a slight concentration dependence for $^2J_{H-F}$ of trifluoromethane in cyclohexane. Gutowsky *et al.* [64] noted a solvent dependence of $^2J_{H-F}$ for trifluormethane which changes from 79.21 Hz in cyclohexane to 79.85 in $C_6F_6$. Both of these results are at variance with the results of Cox and Smith [23] who found $^2J_{H-F}$ of $CHF_3$ and

$CH_2F_2$ to be solvent invariant (Table 5). The values in cyclohexane, $CCl_4$, acetone and DMSO are all smaller than the gas phase value of 79.72 Hz [63)] which is close to the value observed in $C_6F_6$. The abnormal results in $C_6F_6$ are reminiscent of the similar abnormal result observed for $^1J_{29_{Si-F}}$ in the same solvent.

Ihrig [65)] studied a series of rigid bicyclic compounds (5 – 8) obtained from the Diels-Alder reaction of hexachlorocyclopentadiene with fluoroethylenes. In

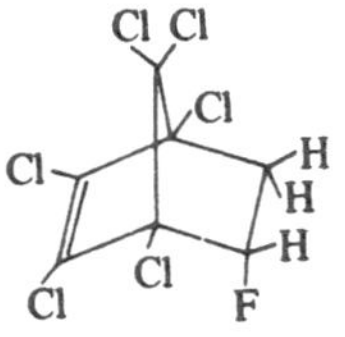

5

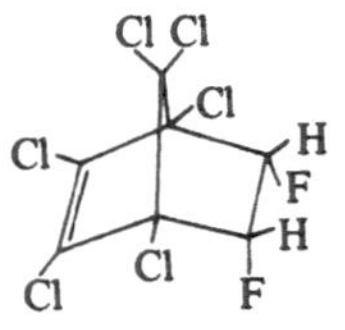

6

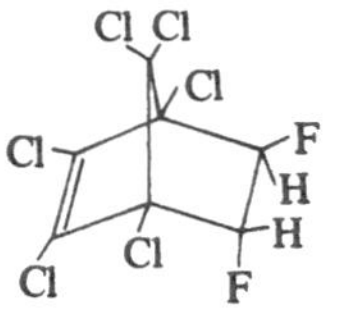

7

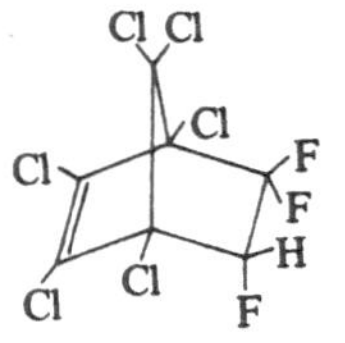

8

every case the geminal H–F coupling constant decreases in solvents of higher dielectric strength (Table 25). The trifluoroethylene adduct, (8) shows the largest change ($\Delta J$ = 2.87 Hz) and the vinyl fluoride adduct shows the smallest change ($\Delta J$ = 0.86 Hz) suggesting that substituent contributions are more important than hybridization. Since all the adducts have approximately the same dipole orientation (away from the H–F group) and since the changes correlate roughly with the solvent dielectric constant a reaction field interaction is postulated. The two geminal H–F coupling constants in the adduct from trans-1,2-difluoroethylene, (7), show essentially the same solvent dependence. This constitutes evidence against any strong specific interactions. Such interaction with either a proton or a fluorine in the exo position ought to be quite different from the same interaction with a proton or a fluorine in the more hindered endo position, leading to different solvent effects for the two coupling constants.

Table 25. *Solvent dependence of H−H and H−F coupling constants in some bicyclic systems*

| Solvent | H−H Couplings | | | H−F Couplings | | |
|---|---|---|---|---|---|---|
| | $^2J_{HH}$ | $^3J_{HH}^{trans}$ | $^3J_{HH}^{cis}$ | $^2J_{HF}$ | $^3J_{HF}^{trans}$ | $^3J_{HF}^{cis}$ |
| Vinyl fluoride Diels-Alder adduct | | | | | | |
| Cyclohexane-d$_{12}$ | −13.33 | 1.83 | 7.19 | 54.37 | 11.92 | 25.01 |
| Benzene-d$_6$ | −13.61 | 1.82 | 7.23 | 54.23 | 12.22 | 25.47 |
| Carbon tetrachloride | −13.41 | 1.85 | 7.22 | 54.31 | 12.01 | 25.10 |
| Deuterochloroform | −13.53 | 1.81 | 7.19 | 54.24 | 12.25 | 25.44 |
| Methylene chloride | −13.66 | 1.79 | 7.20 | 53.90 | 12.34 | 25.66 |
| Acetone-d$_6$ | −13.88 | 1.76 | 7.23 | 53.87 | 12.65 | 25.99 |
| Acetonitrile-d$_3$ | −13.88 | 1.76 | 7.22 | 53.90 | 12.72 | 26.14 |
| Dimethyl sulfoxide-d$_6$ | −13.93 | 1.74 | 7.19 | 53.51 | 13.13 | 26.50 |

*trans*-1,2-Difluoroethylene Diels-Alder adduct

| Solvent | $^3J_{HH}^{trans}$ | Endo $^2J_{HF}$ | Exo $^2J_{HF}$ | Endo $^3J_{HF}^{cis}$ | Exo $^3J_{HF}^{cis}$ |
|---|---|---|---|---|---|
| Cyclohexane-d$_{12}$ | 1.03 | 52.28 | 51.90 | 17.64 | 13.92 |
| Carbon tetrachloride | 1.03 | 52.06 | 51.73 | 17.71 | 13.85 |
| Deuterochloroform | 1.01 | 51.95 | 51.55 | 17.68 | 13.93 |
| Methylene chloride-d$_2$ | 1.02 | 51.75 | 51.44 | 17.70 | 13.97 |
| Acetone-d$_6$ | 1.03 | 51.07 | 50.81 | 17.95 | 14.32 |
| Dimethylsulfoxide-d$_6$ | 0.98 | 50.06 | 49.94 | 18.28 | 14.62 |

*cis*-1,2-Difluoroethylene Diels-Alder adduct

| Solvent | $^3J_{HH}^{cis}$ | $^3J_{HF}^{trans}$ | $^2J_{HF}$ | $^3J_{FF}^{cis}$ |
|---|---|---|---|---|
| Cyclohexane-d$_{12}$ | 6.02 | 1.84 | 52.23 | 16.05 |
| Benzene-d$_6$ | 6.00 | 1.95 | 51.86 | 15.82 |
| Carbon tetrachloride | 6.01 | 1.87 | 52.10 | 16.09 |
| Carbon disulfide | 5.99 | 1.90 | 52.16 | 15.13 |
| Deuterochloroform | 6.01 | 1.89 | 52.00 | 16.05 |
| Tetrahydrofuran | 6.01 | 2.03 | 51.31 | 16.87 |
| Methylene chloride-d$_2$ | 6.00 | 1.99 | 51.86 | 16.39 |
| Cyclopentanone | 6.02 | 2.03 | 51.27 | 16.55 |
| Acetone-d$_6$ | 6.03 | 2.00 | 51.22 | 17.06 |
| Dimethylformamide | 6.02 | 2.08 | 50.96 | 16.75 |
| Acetonitrile-d$_3$ | 6.03 | 2.04 | 51.33 | 17.36 |
| Dimethylsulfoxide-d$_6$ | 6.01 | 2.10 | 50.77 | 15.55 |

Table 25 (continued)

| Solvent | | | |
|---|---|---|---|
| Trifluoroethylene Diels-Alder adduct | $^2J_{HF}$ | $^3J_{HF}^{cis}$ | $^3J_{HF}^{trans}$ |
| Cyclohexane-$d_{12}$ | 51.91 | 9.22 | 0.31 |
| Carbon tetrachloride | 51.72 | 9.13 | 0.31 |
| Deuterochloroform | 51.58 | 9.04 | 0.28 |
| Methylene chloride-$d_2$ | 51.42 | 9.10 | 0.33 |
| Acetone-$d_6$ | 50.26 | 9.24 | 0.33 |
| Dimethylsulfoxide-$d_6$ | 49.04 | 9.36 | 0.27 |

In summary, $^2J_{H-F}$ demonstrates the same pattern of solvent dependence as does $^2J_{H-H}$. However, all the subtleties seem to be enhanced. Usually $^2J_{H-F}$ decreases in solvents of higher dielectric strength, but an appropriate dipole orientation with respect to the H—C—F group can lead to the opposite result as is observed in vinyl fluoride. This situation is perhaps most likely to occur in monofluoro compounds where the fluorine is the principal contributor to the molecular dipole. In either case the electric field effect as postulated with the Pople expression for the contact term produces the correct prediction.

As implied by the dipole orientation effect on the solvent dependence of $^2J_{H-F}$ the reaction field seems to be the major solute-solvent interaction mechanism. However, specific interactions, particularly hydrogen bonding, also are important. Neither finding is particularly surprising since the presence of a fluorine almost inevitably results in the solute molecule being polar.

## 3. $^2J_{F-F}$

Ng, Tang and Sederholm [66] observed the solvent dependence of the F—C—F geminal coupling constant in bromotrifluoroethylene (Table 26). Correlation with solvent dipole moment or specific association were suggested as possible interaction mechanisms. Recognizing that F—F couplings may have significant contributions from coupling mechanisms other than the Fermi contact interaction, the authors suggest modification of "through space" interactions by the solvent.

In a much more extensive and detailed study McDonald and Schaefer [67] reported the solvent dependence of $^2J_{F-F}$ in 1,1-difluoroethylene (Table 27). Ihrig and Smith [62] studied the same compound and obtained identical experimental results for the solvents common to both studies. However, the interpretations given the results in the two studies are different. Both sets of inves-

Table 26. *Solvent dependence of F−F coupling constants in bromotrifluoroethylene*

| Solvent | $^2J_{FF}$ | $^3J_{FF}^{cis}$ | $^3J_{FF}^{trans}$ |
|---------|-----------|------------------|---------------------|
| $CF_2CFBr$ | 73.7 | 56.6 | 122.8 |
| $S{=}C{=}S$ | 71.7 | 56.4 | 123.1 |
| $CF_2BrCF_2Br$ | 73.3 | 56.5 | 123.2 |
| $CF_2ClCFCl_2$ | 73.6 | 56.7 | 123.2 |
| $CFCl_3$ | 73.3 | 56.8 | 123.4 |
| Dioxane | 73.8 | 55.1 | 122.6 |
| $CH_3CSCH_3$ | 74.3 | 53.7 | 122.5 |
| $CH_3COOH$ | 74.3 | 56.0 | 122.7 |
| $(CH_3CH_2)_2O$ | 73.8 | 56.0 | 123.1 |
| $OSCl_2$ | 72.1 | 55.8 | 122.4 |
| $CH_3OH$ | 74.5 | 55.7 | 122.9 |
| $CH_3CH_2OH$ | 73.9 | 55.9 | 122.9 |
| $CH_2{=}CHCH_2Cl$ | 73.1 | 56.1 | 123.0 |
| $CH_3CHO$ | 74.2 | 55.1 | 122.6 |
| $CH_3COCH_3$ | 74.9 | 54.9 | 122.5 |
| $(CH_3CO)_2O$ | 74.4 | 55.4 | 122.6 |
| $CH_3CN$ | 74.6 | 54.9 | 121.9 |

tigators agree that $^2J_{F-F}$ increases ($\Delta J$ = 8.78 Hz) and that there is noticeable evidence of specific interactions in either hydrogen bonding solvents or in solvents having nonbondings electrons. For eleven solvents (excluding those which show evidence of specific interactions) McDonald and Schaefer found a good correlation between $^2J_{F-F}$ and the solvent heat of vaporization at the boiling point. On this basis they suggest dispersion interactions as the major interaction mechanism. Ihrig and Smith found a correlation coefficient of 0.80 between $^2J_{F-F}$ and the reaction field term for a series of fourteen solvents including many of the polar solvents excluded by McDonald and Schaefer. Significantly, $^2J_{F-F}$ increases with increasing reaction field strength, but decreases with increasing dispersion interactions.

Ihrig and Smith extended their study by running a regression analysis including reaction field terms, dispersion terms and various combinations of the solvent refractive index and dielectric constant. The best least squares fit between $^2J_{F-F}$ and solvent parameters was found with a linear function of the reaction field term *and* the dispersion term. The reaction field term was found to be approximately three times as important as the dispersion term and the coefficients of the terms were opposite in sign.

Table 27. *Solvent dependence of coupling constants in 1,1-difluoroethylene (solute concentration < 5 mole %)*

| Solvent | $\epsilon$ | $^2J_{HH}$ | $^3J_{HF}^{cis}$ | $^3J_{HF}^{trans}$ | $^2J_{FF}$ |
|---|---|---|---|---|---|
| Neat | – | −4.69 | 0.75 | 33.83 | 36.49 |
| TMS | 1.91 | −4.64 | 0.61 | 33.79 | 32.21 |
| $C_6H_{12}$ | 2.01 | −4.62 | 0.60 | 33.85 | 30.72 |
| $CS_2$ | 2.61 | −4.70 | 0.73 | 34.17 | 29.24 |
| $CCl_4$ | 2.22 | −4.70 | 0.68 | 34.15 | 31.16 |
| $CBrCl_3$ | 2.38 | −4.70 | 0.67 | 34.19 | 30.19 |
| $CHCl_3$ | 4.63 | −4.84 | 0.77 | 34.46 | 32.32 |
| $HCBrCl_2$ | 4.47 | −4.86 | 0.81 | 34.61 | 31.23 |
| $CHBr_3$ | 4.28 | −4.86 | 0.74 | 34.66 | 28.73 |
| $CH_2Cl_2$ | 8.75 | −4.92 | 0.90 | 34.75 | 34.28 |
| $CH_2ClBr$ | 8.41 | −4.97 | 0.86 | 34.88 | 32.41 |
| $CH_2Br_2$ | 7.04 | −4.93 | 0.86 | 34.93 | 30.75 |
| $C_6H_6$ | 2.26 | −4.98 | 0.86 | 34.67 | 32.83 |
| $C_6H_5F$ | 5.32 | −4.95 | 0.76 | 34.69 | 33.64 |
| $C_6H_5Br$ | 5.33 | −4.91 | 0.81 | 34.72 | 31.45 |
| $C_6H_5NO_2$ | 33.9 | −5.13 | 0.99 | 35.30 | 33.82 |
| p-Dioxane | 2.20 | −5.09 | 1.08 | 35.59 | 34.68 |
| THF | 7.85 | −5.07 | 0.98 | 35.12 | 35.01 |
| $CH_3CO_2H$ | 6.22 | −5.15 | 1.00 | 35.13 | 36.21 |
| $CH_3CO_2CH_3$ | 6.57 | −5.14 | 1.02 | 35.31 | 36.71 |
| $CH_3NO_2$ | 35.9 | −5.26 | 1.13 | 35.65 | 37.51 |
| $CH_3CN$ | 35.8 | −5.25 | 1.10 | 35.56 | 37.48 |
| $CH_3COCH_3$ | 20.2 | −5.24 | 1.08 | 35.53 | 36.71 |
| DMF | 35.8 | −5.38 | 1.22 | 36.16 | 36.82 |
| DMSO | 47 | −5.47 | 1.29 | 36.66 | 35.64 |

The difference in $\Delta J$ for bromotrifluoroethylene and 1,1-difluoroethylene (9.78 Hz *vs* 1.8 Hz) may be significant. Certainly 1,1-difluoroethylene would be expected to have ᴀ much larger net dipole and, if reaction field interactions are important, to exhibit the larger effect.

Since the geminal F−F coupling constant is positive and the dipole moment is oriented towards the fluorines the observed increase in $^2J_{F-F}$ is expected from the simple contact term variation postulated earlier for $^2J_{H-H}$ and $^2J_{H-F}$. The situation is entirely analogous to that of $^2J_{H-F}$ in vinyl fluoride.

Ng, Tang and Sederholm [66] also reported the solvent dependence of $^2J_{F-F}$ in the $sp^3$ hybridized compound $CF_2BrCFBr_2$ (Table 28). Again, the coupling constant increases and there is an apparent correlation between solvent polarity and the magnitude of the change.

Table 28. *Solvent dependence of $^2J_{F\text{-}F}$ in $CF_2BrCFBr_2$*

| Solvent | $^2J_{FF}$ |
|---|---|
| $CS_2$ | 165.0 |
| $CF_2BrCFBrCl$ | 166.6 |
| $(CH_3CH_2)_2O$ | 169.3 |
| $CH_3COOH$ | 169.7 |
| $CH_3CH_2OH$ | 169.8 |
| 1,4-Dioxane | 170.7 |
| $(CH_3CO)_2O$ | 170.7 |
| $CH_3OH$ | 170.9 |
| $CH_3CN$ | 171.2 |
| $CH_3COCH_3$ | 171.5 |

As an example of the different combinations arising as couplings over several bonds are studied we note in passing that $^2J_{F\text{-}F}$ across chlorine in the T shaped molecule $ClF_3$ varies from 444 Hz in the gaseous state to 420 Hz in $CCl_4$ and 426.3 Hz in $CCl_3F$ [68].

### 4. $^2J_{M\text{-}H}$; $M = {}^{119}Sn, {}^{117}Sn, {}^{207}Pb, {}^{199}Hg$

Laszlo and Speert [31] provided the best evidence now available for the effect of dispersion interactions on coupling constants by studying the solvent dependence of $^2J_{{}^{119}Sn\text{-}H}$, $^2J_{{}^{117}Sn\text{-}H}$ and $^2J_{{}^{207}Pb\text{-}H}$ in the centrosymmetric molecules tetramethyltin and tetramethyllead. Since these molecules have no dipole moments they do not produce a reaction field. There is no evidence for significant specific interactions such as hydrogen bonding. Thus, the dispersion interactions constitute the major mechanism by which these molecules can interact with their surroundings. (Stark effects are also a possibility.) As shown in Table 29 all three coupling constants change by 1.7–2.2 Hz. Superficially it appears that all three couplings increase in solvents of higher dielectric constant. However, careful examination reveals a number of discrepancies. A correlation between a decrease in the coupling and an increase in the refractive index is as good or better. Plots of the coupling constants *vs* either the McRae term (dispersion interactions) or the reaction field term show correlation coefficients in the vicinity of 0.7.

The dispersion interaction is nicely isolated from other factors by examining the solutes in mixtures of heptane and carbon disulfide; nonpolar solvents of similar dielectric constant, but quite different refractive index. Under these conditions there is a better than 0.95 correlation between $^2J_{M\text{-}H}$ and the McRae term as illustrated for $^2J_{{}^{119}Sn\text{-}H}$ in Fig. 7.

167

Table 29. *Solvent dependence of* $^2J_{M\text{-}H}$ *in tetramethyltin and tetramethyllead*

| Solvent | $\epsilon$ | $n$ | $J_{119_{Sn\text{-}C\text{-}H}}$ | $J_{117_{Sn\text{-}C\text{-}H}}$ | $J_{207_{Pb\text{-}C\text{-}H}}$ |
|---|---|---|---|---|---|
| Cyclohexane | 2.02 | 1.4262 | 54.04 | 51.68 | – |
| 1,4-Dioxane | 2.21 | 1.4232 | 54.78 | 52.38 | 62.44 |
| Carbon tetrachloride | 2.23 | 1.4630 | 53.99 | 51.69 | 61.43 |
| Benzene | 2.27 | 1.5011 | 54.22 | 51.90 | 61.50 |
| Carbon disulfide | 2.64 | 1.6255 | 53.63 | 51.30 | 60.72 |
| Deuterochloroform | 4.70 | 1.4450 | 54.27 | 51.98 | 61.97 |
| Aniline | 6.98 | 1.5863 | 54.16 | 51.63 | 61.24 |
| Methylene chloride | 8.9 | 1.4237 | 54.66 | 52.24 | 62.78 |
| Pyridine | 12.3 | 1.5095 | 54.31 | 51.99 | 61.73 |
| Benzaldehyde | 17.8 | 1.5446 | 54.32 | 51.92 | 61.95 |
| Acetone-$d_6$ | 20.5 | 1.3592 | 54.96 | 52.56 | 63.25 |
| Benzonitrile | 25.2 | 1.5289 | 54.58 | 52.26 | 61.77 |
| Hexamethyl-phos-phoric triamide | 29.6 | 1.4584 | 54.25 | 52.12 | 62.08 |
| Methanol | 32.6 | 1.3288 | 54.62 | 52.34 | 62.74 |
| Nitrobenzene | 34.6 | 1.5562 | 54.58 | 52.16 | 61.74 |
| Dimethyl formamide | 36.7 | 1.4319 | 54.81 | 52.69 | 62.94 |
| Acetonitrile | 37.5 | 1.3442 | 55.55 | 53.06 | 63.91 |
| Dimethyl sulfoxide | 48.9 | 1.4787 | 54.83 | 52.45 | 62.24 |

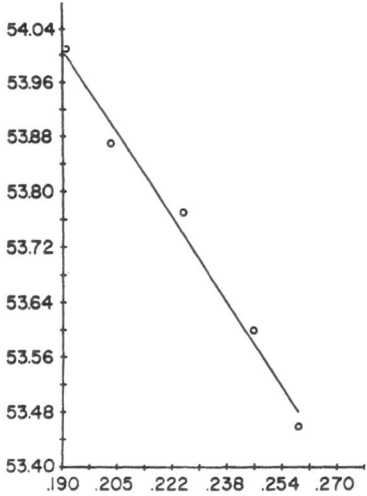

Fig. 7. Plot of $J_{119_{Sn\text{-}C\text{-}H}}$ [ordinate] versus McRae's term, $(n^2-1)/(2n^2+1)$ [abscissa], for $n$-heptane-carbon disulfide mixtures

The Laszlo results raise serious questions about the correlations suggested between the signs of coupling constants and the direction of the solvent induced change of a coupling constant. The reduced coupling constants $^2J_{Sn-H}$ and $^2J_{Pb-H}$ are both negative, yet they *increase* in solvents of higher refractive index. This is opposite to the trend generally observed for geminal H—H and H—F coupling constants which appear to *decrease* in solvents of increasing dielectric constant. This observation may well indicate that dispersion interactions and reaction field interactions produce opposite effects on coupling constants. Such a situation is consistent with the results of Coyle *et al.* [29] for $^1J_{Si-F}$ (Table 10) which displays behavior opposite to that of most other single bond couplings. It is supported by the report of Ihrig and Smith [62] who suggest a combination of reaction field and dispersion interactions *which have opposite effects* to explain the solvent dependence of coupling constants in fluoroethylenes. However, such a conclusion is not consistent with the suggested dispersion interaction effects on $^2J_{F-F}$ in 1,1-difluoroethylene [67] or with Laszlo's report [31] that the increase in $^1J_{C-H}$ for $CH_2Cl_2$ and $CH_2Br_2$ correlates better than 0.96 with the McRae dispersion term.

Other alternatives are possible. Perhaps present interpretations of the mechanism for solvent effects on coupling constants may be seriously deficient or completely wrong. Alternatively, couplings between heavy atoms such as tin and lead may involve factors not present for the lighter elements.

Only one other clear case of a solvent dependent $^2J_{M-H}$ coupling constant has been reported. Hatton *et al.* [69] report the solvent dependence of $^2J_{Hg-C-H}$

Table 30. *Solvent dependence of $^2J_{Hg-H}$ in monomethyl- and monoethylmercury compounds*

| Compound (conc. 5 mole %) | | Solvent | | |
|---|---|---|---|---|
| | Cyclohexane | Benzene | Pyridine | $D_2O$ |
| $CH_3-Hg-CH_3$ | 100.6 | | 104.3 | |
| $CH_3-Hg-CN$ | | 176.0 | 178.0 | |
| $CH_3-Hg-OH$ | | 204.0 | 214.2 | |
| $(CH_3Hg)_2SO_4$ | | 205.0 | 216.0 | |
| $(CH_3Hg)_2C_2O_4$ | | 205.0 | 215.2 | |
| $CH_3HgOAC$ | | 214.3 | 220.8 | 233.4 |
| $(CH_3Hg)_3PO_4$ | | – | 220.5 | 233.2 |
| $CH_3HgNO_2$ | | 240.6 | 227.0 | 259.2 |
| $CH_3HgClO_4$ | | 259.8 | 233.2 | 259.6 |
| $C_2H_5HgCN$ | $J_{CH_2}$ | 182.0 | 186.0 | |
| | $J_{CH_3}$ | 222.0 | 222.0 | |
| $C_2H_5HgNO_3$ | $J_{CH_2}$ | 236.6 | 233.0 | 250.0 |
| | $J_{CH_3}$ | 348.0 | 311.0 | 369.0 |

in a variety of compounds of the form $CH_3HgX$ and $CH_3CH_2HgX$ (Table 30). The effect appears to be largest in those compounds which exhibit the least tendency to ionize. The data are insufficient for any detailed analysis or comment.

## 5. Two-Bond Couplings to Phosphorus

Surprisingly little data is available concerning the solvent dependence of two-bond phosphorus couplings. Gordon and Griffin [70] reported the solvent dependence of $^2J_{P-C-H}$ in three benzylphosphonium salts (Table 31). The coupling constants were independent of concentration and of the anion (Cl⁻, Br⁻, I⁻). Specific interactions from donor solvents such as DMSO or possibly reaction field effects were suggested as causative factors. Similar results have been reported by Martin and Besnard [71] for $CH_3POCl_2$ in a limited range of nonpolar solvents. Gordon and Griffin specifically note that the geminal P–C–H coupling constant in benzyl phosphonates is *not* solvent dependent.

Table 31. *Solvent dependence of the P–C–H coupling constant in benzylphosphonium salts*

| Solvent | $|J_{PCH}|$ (Hz) |
|---|---|
| $[C_6H_5)_3PCH_2C_6H_5]^+$ | |
| $CF_3COOH$ | 14.0 |
| $CDCl_3$ | 14.1 |
| $CH_3CN$ | 14.6 |
| $CH_3COOH$ | 15.0 |
| $(CH_3)_2NCHO$ | 15.5 |
| $(CH_3)_2SO$ | 15.6 |
| $[n\text{-}C_4H_9)_3PCH_2C_6H_5]^+$ | |
| $CF_3COOH$ | 13.8 |
| $(CH_3)_2SO$ | 15.3 |
| $CDCl_3$ | 15.4 |
| $CH_3CN$ | 15.6 |
| $[(C_6H_5)_3PCH_2C_6H_4COOCH_3)\text{-}p]^+$ | |
| $CF_3COOH$ | 14.7 |
| $CDCl_3$ | 15.2 |
| $(CH_3)_2SO$ | 16.2 |

Fields, Green and Jones [33] studied the solvent dependence of all the coupling constants in bis(trifluoromethyl)phosphine and noted a decrease of $^2J_{P-C-F}$ with increasing dielectric constant of the solvent (Table 14). This does not result from an increase in coordination at phosphorus since such effects are known to cause a drastic increase in the P—C—F coupling constant. Field *et al.* suggest hydrogen bonding interactions in accord with the effects proposed by Evans [18] (Table 1) for $^1J_{C-H}$ in chloroform and note that there is a good linear relationship (confidence limit > 99.9 %) between $^1J_{C-H}$ and $^2J_{P-C-H}$. Since $^2J_{P-C-F}$ is positive in bis(trifluoromethyl)phosphine the decrease in solvents of higher dielectric constant or upon hydrogen bonding is in accord with the idea that geminal coupling constants decrease with increasing reaction field or hydrogen bonding.

## 6. Two-Bond Couplings to Nitrogen

As was the case with phosphorus couplings, very little has been reported concerning the solvent dependence of geminal coupling constants involving nitrogen. Alger and Gutowsky [72] used spin-echo techniques to examine the solvent dependence of $^2J_{^{14}N-C-F}$ in 2-fluoropyridine (Table 32). The experimental problems involved in this work were considerable, but within the rather large experimental error ($\pm$ 1 Hz) there was a linear relationship with either reaction field or Stark terms.

Table 32. *Solvent dependence of* $^2J_{N-C-F}$ *in 2-fluoropyridine*

| Solvent | $J_{NF}$ (Hz) |
|---|---|
| Carbon tetrachloride | 48.7 |
| Thiophene | 47.6 |
| Pyridine | 46.6 |
| Neat | 45.8 |
| Ethanol | 45.0 |
| Methanol | 44.5 |
| Formamide | 44.0 |

Crépaux *et al.* [73,74] conducted a more extensive study concerning the solvent dependence of $^2J_{^{15}N-C-H}$ in quinoline and in a series of oximes of the general form

$$\begin{array}{c} R_1 \\ {\Large\diagdown} \\ {\phantom{R}} \end{array} C{=}N {\diagdown}{\phantom{OH}} OH$$

$$\begin{array}{c} {\phantom{R}} \\ {\Large\diagup} \\ R_2 \end{array}$$

Their results are listed in Table 33. Signs given are the absolute signs of the reduced coupling constants. It is notworthy that the *anti* and *syn* couplings are of opposite absolute sign and both *increase* (in the absolute sense) in solvents of higher dielectric constant. Addition of strong acid produces the

171

Table 33. *Solvent dependence of* $^2J_{15_{N-C-H}}$ *in quinoline and selected oximes*

| $R_1$ | $R_2$ | Solvent | $^2J_{NH}$ (anti) | $^2J_{NH}$ (syn) |
|---|---|---|---|---|
| H | H | $iPr_2O$ | $\pm 14.0_0$ | $-2.5_3$ |
| | | $Et_2O$ | $13.9_1$ | $-2.4_3$ |
| | | $CH_3CN$ | $14.1_9$ | $-2.2_7$ |
| H | $CH_3CH_2-$ | Pentane | $15.5_0$ | $-$ |
| | | $CH_2Cl_2$ | $16.2_0$ | $-$ |
| | | $CH_3CN$ | $16.7_5$ | $-$ |
| $CH_3CH_2-$ | H | Pentane | $-$ | $-3.0_1$ |
| | | $CH_2Cl_2$ | $-$ | $-2.9_5$ |
| | | $CH_3CN$ | $-$ | $-2.6_8$ |
| $(CH_3)_3C-$ | H | Pentane | $-$ | $-3.0_0$ |
| | | $CH_2Cl_2$ | $-$ | $-2.7_8$ |
| | | $(CH_3)_2SO$ | $-$ | $-2.7_8$ |
| Quinoline | | Pentane | $-$ | $+10.7_8$ |
| | | $CH_3CN$ | $-$ | $+11.0_0$ |

same effect. The initial suggestion that this behavior was opposite to that observed for geminal H–H coupling constants may be reinterpreted in terms of the reaction field and dipole orientation effect. Whereas for most H–C–H groups the dipole moment is oriented away from the hydrogens, in the oximes and quinolines the dipole is oriented from carbon to nitrogen, effectively towards the hydrogen and nitrogen. The situation is practically identical to that found in vinyl fluoride and, as predicted, the coupling constants change in the opposite direction.

Paolillo and Becker [37] noted the solvent dependence of $^2J_{15_{N-C-H}}$ involving the $sp^3$ hybridized nitrogen in trimethylamine. Values of 0.85, 0.80 and 0.60 Hz in cyclohexane, deuterochloroform and methanol respectively probably represent a real change, but the data do not permit any detailed analysis.

## 7. Two-Bond Couplings to Carbon

Bell and Danyluk [26] reported the solvent dependence of $^2J_{13_{C-C-H}}$ and $^2J_{13_{C-C-H}}$ in the 1-fluoro-1,2-dichloroethylenes (Table 9). These limited results seem to be in accord with generalizations concerning other solvent dependent two-bond coupling constants. The data is insufficient to permit extensive comment.

## 8. Summary of Solvent Effects on Two-Bond Couplings

It appears clear that all geminal coupling constants should normally be expected to be solvent dependent. The magnitude of the expected changes is significant. While the numerical changes observed for H—C—H couplings (0.2 – 2.4 Hz) are small they frequently correspond to differences of 5, 25, 50 or even 85 % of the value of the coupling constant. Generally, the smaller values (0.2 – 0.8 Hz) are observed for $sp^3$ hybridized carbon systems which do not have heteroatoms bound to the carbon of the H—C—H group. Conversely, the larger values are found with $sp^2$ hybridized systems and/or when heteroatoms are bound to the $CH_2$ group. Changes in coupling constants involving heavier atoms are numerically larger (1 – 20 Hz), but since couplings involving heavy atoms are usually large the changes correspond to only 1 – 10% of the value of the coupling. The geminal F—C—F coupling which changes by 32% is a notable exception. The few cases in which a coupling has been studied and found not to be solvent dependent must represent either exceptional cases or situations where the solvents selected were sufficiently similar to the solute in dielectric constant and refractive index that the changes were too small to be readily observed.

It further appears that geminal coupling constants decrease in the absolute sense for reaction field interactions, given a "normal" dipole orientation. Abnormal or opposite dipole orientation produces the opposite change in the coupling constant.

Normal orientation     Abnormal orientation

Fig. 8. Normal and abnormal dipole orientation for determining solvent effects on $^2J_{A-B}$

The definition of normal and abnormal proposed here is arbitrary, but convenient. The definition of normal is the most common situation encountered where one or both of the coupled nuclei is hydrogen in an organic molecule. It thus appears that absolute signs of any two-bond coupling constant between any nuclei can be determined from the solvent dependence of the coupling subject to three conditions: (1) the solute molecule must be polar so that reaction field interactions predominate; (2) due attention must be paid to the dipole orientation with respect to the atoms in question, and (3) a sufficient number of solvents of varying dielectric constant must be used to avoid the occasional spurious result. In connection with condition two it should be noted that the results for $^2J_{H-F}$ in vinyl fluoride and cis-12,-difluoroethylene, and $^2J_{N-C-H}$ in the oximes indicate that a dipole directed along one of the two bonds connecting the coupled nuclei is effectively an abnormal dipole orientation.

As implied above, the major interaction mechanism affecting geminal couplings seems to be the reaction field. Dispersion interactions also play a role and their effects are more important for nonpolar solutes and with solvents having non-bonding electrons or solvents which are heavily halogenated. Specific interactions, particularly hydrogen bonding, are also important. Hydrogen bonding effects are particularly noticeable with couplings involving atoms having nonbonded electrons such as nitrogen and fluorine.

## D. Three- and Four-Bond Couplings

Three-bond coupling constants present even more problems concerning the question of solvent dependence than do the two-bond coupling constants. Most important of these is the fact that three-bond coupling constants are known to vary with the dihedral angle between the C—H bonds. In turn, the dihedral angle can change as a function of molecular conformation which is frequently solvent dependent. In this connection it must be recognized that vicinal couplings may change as a result of conformational changes in an adjacent part of the molecule even though that portion of the molecule involving the vicinal coupling constant is conformationally invariant. We will be concerned with solvent induced changes of vicinal coupling constants which occur in rigid systems or in addition to (or in spite of) solvent induced conformational changes. The net result of these problems plus increased possibilities in terms of the nature of the intervening atoms, hybridization, etc., is that relatively little is known concerning the solvent dependence of three- and four-bond coupling constants and the data which is available does not demonstrate any straightforward patterns as is the case for one- and two-bond coupling constants.

### 1. $^3J_{H\text{-}C\text{-}C\text{-}H}$

There is some question as to whether vicinal H—H coupling constants are or are not solvent dependent. Most of the studies reported earlier concerning the solvent dependence of geminal H—H coupling constants also reported that the vicinal H—H coupling constants in the same molecules are solvent invariant within experimental error (generally ±0.1 Hz or better). Thus, no solvent dependent vicinal H—H coupling constants were found in any of the studies conducted on mono- and disubstituted olefins. Yet, Laszlo and Bos [75] report that $^3J_{H\text{-}H}^{cis}$ of ClCH=CHOEt varies from 6.3 Hz in cyclohexane to 4.2 Hz in DMSO! Laszlo has subsequently questioned these results [2]. Cox and Smith [56] during their study of three-membered rings found some significant changes in vicinal H—H couplings and some which were essentially solvent invariant (Table 34).

Table 34. *Summary of solvent induced changes of coupling constants in three-membered rings:* $\Delta J = J_{C_6H_{12}} - J_{DMSO}$

| Compound | $\Delta J_{gem}$ | $\Delta J_{trans}$ | $\Delta J_{cis}$ |
|---|---|---|---|
| Styrene | −0.69 | 0.18 | 0.24 |
| Styrene sulfide | −0.46 | 0.46 | 0.20 |
| 2,2-Dichlorocyclopropylbenzene | −0.34 | 0.50 | 0.52 |

Laszlo and Bos observed a decrease in solvents of increasing polarity. Cox and Smith observed an increase. All the couplings are positive.

Hutton and Schaefer [76] and Smith and Ihrig [77] report changes of 0.14 ± 0.1 Hz for the *ortho, meta* and/or *para* coupling constants of *p*-nitroanisole, two dinitrotoluenes and 2,4-dinitrochlorobenzene. Both reports indicate that other aromatic compounds investigated showed even less solvent dependence.

Compounds in which the carbons are $sp^3$ hybridized display the same confused situation. Erickson [78] reports that the *trans* vicinal H–H coupling constant of *dl*-dibromosuccinic anhydride (which is reasonably rigid) varies from 2.5 Hz in $CHCl_3$ to 6.0 Hz in acetone and dioxane. The same paper reports that the *meso* dibromosuccinic anhydride and the two corresponding dichloro compounds do not display any solvent dependence of their coupling constants. (Erickson also reports that $^1J_{C-H}$ of the *dl*- dibromide *decreases* from 172 Hz in chloroform to 166 Hz in acetone and 165 Hz in dioxane; exactly the opposite behavior from that observed for any other $^1J_{C-H}$ coupling ever studied). It is at least possible that these data result from chemical degradation of the solute rather than from true solvent effects as discussed here.

Smith and Cox [57] found the geminal H–H coupling constants in a series of rigid bicycloheptenes to vary with solvent, but report that both the *cis* and *trans* vicinal couplings (dihedral angle = 0° and 120° respectively) are solvent invariant.

Finegold [58] examined ethylene-2,1,3-thiadioxalone-2 *(9a)*, the corresponding propylene 3,1,2-thiadioxalones *(9 b, c)* and propylene-3,1,2-dioxalone *(9 d)* in a limited number of solvents. In addition to the expected decrease in the geminal coupling, the vicinal *gauche* coupling of *(9 a)* is reported to vary by 0.76 Hz (Table 35). With the exception of benzene there is a general correspondence between both the decrease in $^3J_{H-H}$ and the decrease in $^2J_{H-H}$ with increasing dielectric constants of the solvent. A similar, but smaller (0.22 Hz) change is noted for the *trans* vicinal coupling constant.

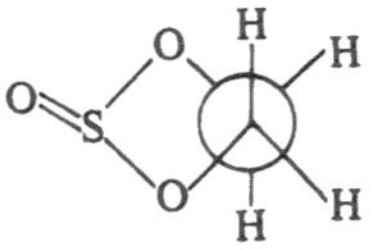

*9a*

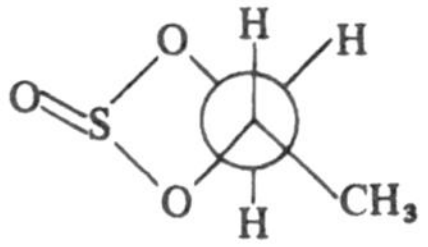

*9b*

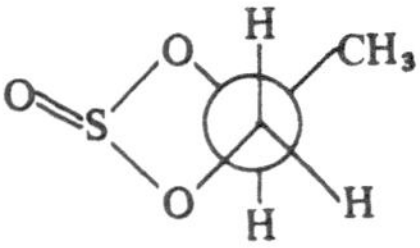

*9c*

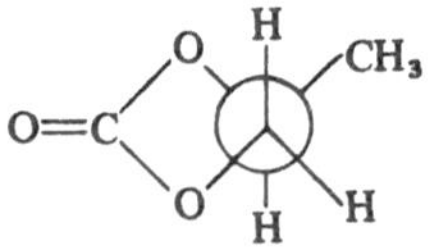

*9d*

Table 35. *Solvent dependence of coupling constants for ethylene-3,1,2-thiadioxalone-2 in various solvents*

| Solvent | $^2J_{H-H}$ | $^3J_{H-H}^{cis}$ | $^3J_{H-H}^{trans}$ |
|---|---|---|---|
| Cyclohexane | − 8.10 | 7.00 | 6.48 |
| Carbon tetrachloride | − 8.15 | 7.00 | 6.58 |
| Dioxane | − 8.31 | 6.83 | 6.60 |
| Benzene | − 8.25 | 7.46 | 6.56 |
| Deuterochloroform | − 8.42 | 6.88 | 6.70 |
| Pyridine | − 8.31 | 6.74 | 6.52 |
| Acetone-$d_6$ | − 8.29 | 6.77 | 6.51 |
| Neat | − 8.41 | 6.70 | 6.52 |
| Nitrobenzene | − 8.42 | 6.76 | 6.54 |
| Acetonitrile | − 8.42 | 6.70 | 6.58 |

One of the propylene derivatives displays the same effects to a lesser degree. The propylene carbonate coupling constants are solvent invariant.

Finegold recognizes that the apparent changes in the vicinal coupling constants might arise from conformation changes, but presents reasonable arguments against such events. It is also possible that the observed variations are an artifact of the spectral analysis technique. An $A_2B_2$ approach was used, although the ethylene derivative, if rigid and staggered, is really an ABCD system. Since the changes in the vicinal coupling parallel the changes in the geminal coupling which in turn are in accord with numerous other studies the solvent dependence of the vicinal coupling constant is probably real.

In summary, a few three-bond H−C−C−H coupling constants have been shown to be solvent dependent. Many others have been shown to be solvent invariant. Generally the change, if any, is small, on the order of 0.1 − 0.3 Hz. However the few cases which seem well documented present a clear *caveat;* utilization of changes in vicinal coupling constants as a measure of conformational changes, energy barriers, etc., may lead to serious errors unless it is unambiguously established that the observed changes are not the result of intrinsic solvent effects.

## 2. $^3J_{H-C-C-F}$

Vicinal H−F coupling constants have received surprisingly detailed study, especially in olefins. Particularily useful are the studies of Smith and Ihrig [61,62] and of MacDonald and Schaefer [67] on the related series, vinyl fluoride, *cis*- and *trans*-1,2-difluoroethylene, 1,1-difluoroethylene and trifluoroethylene (Tables 23 and 27). The vicinal H−F couplings in this series are positive and increase with increasing polarity of the solvent, with the single exception of the *cis* coupling in trifluoroethylene which is negative and displays little or no change ($\Delta J = 0.16 \pm 0.1$ Hz). A number of regular trends are apparent. For all the vicinal coupling constants in this series there is an excellent correlation between $\Delta J$ and $\sqrt{J}$. Thus, since the magnitude of the vicinal H−F couplings decreases with increasing substitution of halogen for hydrogen, the $\Delta J$ values show a corresponding decrease (e.g. for the *trans* coupling from 7.57 Hz in vinyl fluoride to 1.80 Hz in trifluoroethylene). Similarily, for any given compound $\Delta J$ is larger for the large *trans* coupling and smaller for the smaller *cis* coupling with one exception; the change of 0.69 Hz observed in the abnormally small (0.60 Hz) *cis* coupling in 1,1-difluoroethylene. Unlike the geminal H−F and H−H couplings there is no evidence of any dipole orientation dependence for the solvent effect on vicinal H−F couplings.

Particularly startling is the 0.77 Hz change observed for the *cis* H−F coupling constant of *trans*-1,2-difluoroethylene. This compound does not have any net dipole moment and the geminal H−F coupling does not display any appreciable solvent dependence, yet the vicinal coupling involving the same nuclei is solvent dependent.

Turning to the solute-solvent interaction question the most striking result is the 0.9 or better correlations between *any* two vicinal coupling constants in the same or different molecules as a function of solvent [62]. This observation clearly indicates that whatever is happening, it is the same for all molecules in the series. Ihrig and Smith report correlation coefficients of 0.8 or better between the coupling constants and the reaction field term of all vicinal H−F coupling constants. If associating solvents such as benzene, DMSO and trifluoroacetic acid are excluded from the analysis the correlation coefficients are usually 0.9 or better. In line with earlier suggestions that both reaction field interactions and

177

dispersion interactions might affect coupling constants a regression analysis was performed. In several cases, particularly the *trans*-1,2-difluoroethylene results, the best fit to a linear least squares regression is obtained with *both* the dispersion and the reaction field interactions included.

The data of Bell and Danyluk [26] for *cis*- and *trans*-1,2-dichlorofluoroethylenes reported earlier (Table 9) display results completely in accord with the trends noted for the fluoroethylenes. The results of Hutton and Schaefer [79] for 1-chloro-1-fluoroethylene (Table 36) are somewhat different. Contrary to the

Table 36. *Solvent dependence of H−H and H−F coupling constants of 1-chloro-1-fluoroethylene*

| Solvent | $^2J_{HH}$ | $^3J_{HF}^{cis}$ | $^3J_{HF}^{trans}$ |
|---------|-----------|------------------|--------------------|
| Neat | 4.02±0.04 | 9.1 | 36.9 |
| TMS | 3.83±0.05 | 7.7 | 37.7 |
| $C_6H_{12}$ | 3.78±0.04 | 7.7 | 37.7 |
| $CCl_4$ | 3.87±0.07 | 7.5 | 37.9 |
| $C_6H_6$ | 4.24±0.05 | 8.2 | 38.8 |
| $CBrCl_3$ | 3.97±0.06 | 7.5 | 38.3 |
| $CHBr_3$ | 4.06±0.04 | 8.1 | 38.6 |
| $CHCl_3$ | 4.06±0.05 | 8.9 | 37.5 |
| $CH_2ClBr$ | 4.19±0.03 | 9.5 | 37.6 |
| $CH_2Cl_2$ | 4.25±0.05 | 9.6 | 37.7 |
| Acetone | 4.58±0.04 | 11.1 | 37.3 |
| $CH_3CN$ | 4.54±0.05 | 11.3 | 37.3 |
| DMF | 4.68±0.05 | 11.7 | 37.6 |
| DMSO | 4.76±0.05 | 12.3 | 37.9 |

above results, the *cis* coupling shows a  much larger change (4.8 Hz) than the *trans* coupling (0.5 Hz). Why is not at all clear. Also, plots of the coupling constants *vs* the reaction field term display considerable curvature which is removed in the case of the *cis* coupling by plotting against $\epsilon^{1/2}$. The trans H−F coupling shows a fair correlation with dispersion interactions as evidenced by a reasonably linear plot against heats of vaporization of the solvents.

Kumar [80] has reported 10% increases in the *ortho* and *meta* H−F coupling constants of 1,3-difluoro-4,6-dinitrobenzene (Table 37). The couplings also vary with concentration. Since the changes correlate roughly with the dielectric constant of the solvent, reaction field interactions are suggested as the causative factor. Hutton *et al.* [81] report similar changes for the *ortho* H−F coupling constants in 3-chloro- and 3-bromo-4,6-dinitrofluorbenzene ($\Delta J = 0.8$ Hz). The *meta* couplings of these latter compounds, while probably solvent dependent display only very small ($\sim 0.1$ Hz) changes. In all cases studied to date the *para* H–F coupling constants seem to be solvent invariant.

Table 37. *Solvent dependence of ortho and meta H–F coupling constants of 1,3-difluoro-4,6-dinitrobenzene*

| Solvent | $J_{HF^m}$ | $J_{HF^0}$ |
|---|---|---|
| Benzene | 7.00±0.08 | 10.01±0.04 |
| Deuterochoroform | 7.49±0.05 | 9.69±0.07 |
| Dichloromethane | 7.61±0.04 | 9.93±0.05 |
| Acetone | 7.66±0.05 | 10.61±0.05 |
| Acetonitrile | 7.70±0.04 | 10.53±0.05 |
| Tetrahydrofuran | 7.65±0.04 | 10.58±0.05 |
| Dioxane | 7.73±0.05 | 10.61±0.05 |

3. $^3J_{F-C-C-F}$

The solvent dependence of three-bond F–F couplings has not been extensively studied. Where it has been reported it is usually as a part of the study of some other coupling constant. The single exception is the study of Ng, Tang, and Sederholm [66] on bromotrifluoroethylene (Table 26). The *cis* coupling changes from 56.8 Hz in CFCl$_3$ to 53.7 Hz in thioacetone. The *trans* coupling changes from 123.4 in CFCl$_3$ to 121.9 Hz in acetonitrile. Contrary to the implication presented by the solvents in which the extreme values are observed, there is no particular correlation between the solvent polarity and the direction or magnitude of the solvent induced change in the coupling constant. The authors suggest that specific interactions or moderation of through-space coupling by the solvent might be the interaction mechanism. This suggestion rests primarily on the difference in the magnitude and temperature dependence of the *cis* and *trans* coupling constants. The conclusion is somewhat more tenuous in light of more recent studies on other coupling constants.

Ihrig and Smith [62] report similar changes for $^3J_{F-F}^{cis}$ of *cis*-1,2-difluoroethylene and $^3J_{F-F}^{trans}$ of *trans*-1,2-difluoroethylene (Table 23). The *cis* coupling changes by 2.82 Hz from 17.77 Hz in trifluoroacetic acid to 20.59 Hz in DMSO. The *trans* coupling changes by 3.59 Hz from –133.79 in CS$_2$ to –130.20 Hz in acetonitrile. Again there is no particular correlation between solvent polarity and the direction or magnitude of the observed changes. The *cis* coupling constant was found to show a correlation with the McRae dispersion term (0.75). The *trans* coupling showed correlations of 0.83 with the reaction field term and 0.79 with the dispersion term. There was also a noticeable correlation between solvent

179

shape and the magnitude of the *cis* F–F coupling constant. Similar solvent effects on the ortho F–F coupling constant of iodo- and bromofluorobenzene are reported by Cooper [82].

Gutowsky *et al.* [64] presented the only example of a solvent dependent three-bond F–F coupling constant involving an $sp^3$ hybridized carbon. They found the coupling between the $CF^3$ group and the fluorines in the 2 and 6 positions of perfluorotoluene to vary from 22.00 Hz in $CS_2$ to 22.70 Hz in $C_6F_6$. Studies of the temperature dependence of the coupling and other considerations led them to the conclusion that the observed change is a direct solvent effect and not the result of conformation changes, variations in the $CF_3$ rotational rate and the like.

## 4. Other Solvent Dependent Three- and Four-Bond Couplings

Crepaux *et al.* [74] report a noticeable solvent dependence of the three-bond $^{15}N=N-CH$ coupling for several oximes and for quinoline (Table 38). No obvious trends are found.

Table 38. *Solvent dependence of* $^3J_{15_{N=C\text{-}C\text{-}H}}$ *in oximes and quinoline (Signs are those of the reduced coupling)*

| Solvent | Anti coupling | Syn coupling |
| --- | --- | --- |
| $(CH_3)_2C=N-OH$ | | |
| $H_2O$ | (+)4.0 | (+)2.2 |
| $C_6H_6$ | 4.0 | 1.8 |
| $CS_2$ | 3.7 | 2.0 |
| $CF_3COOD$ | 4.5 | 3.2 |
| $(CH_3)C-C=N-OH$ | | |
| $\quad\mid$ | | |
| $CH_3$ | | |
| $C_6H_6$ | | (+)1.8 |
| $CS_2$ | | 1.8 |
| HCOOH | | 3.2 |
| $CF_3COOD$ | | 3.3 |
| $D_2SO_4$ | | 3.4 |
| Quinoline | | |
| $CCl_4$ | (+)1.3 | |
| $CF_3COOD$ or $D_2SO_4$ | 4.5 | |

Rader [83] reported the solvent dependence of H–C–O–H of methanol at dilute concentration in several solvents (Table 39). The effect is small, but real.

No particular relationship between solvent polarity and the observed change is apparent. Rader notes that the experimental conditions are such that no appreciable intramolecular hydrogen-bonding occurs between methanol molecules, presumably ruling out the possibility that the change arises from solvent induced variation of the self-association equilibrium. Differences in solvent-solute association are still a possible interaction mechanism.

Table 39. *Solvent dependence of* $^3J_{H\text{-}C\text{-}O\text{-}H}$ *of methanol*

| Solvent | Concn, M | $J_{H\text{-}C\text{-}O\text{-}H}$, Hz |
|---------|----------|----------------------------------------|
| Neat | | 5.18±.05 |
| Dimethyl sulfoxide | 3.00 | 5.16 |
| | 0.50 | 5.14 |
| | 0.100 | 5.21 |
| | 0.030 | 5.23 |
| Tetramethylurea | 3.00 | 5.17 |
| | 0.50 | 5.22 |
| | 0.100 | 5.20 |
| | 0.030 | 5.22 |
| Carbon tetrachloride | 0.50 | 5.31 |
| | 0.100 | 5.34 |
| | 0.030 | 5.37 |
| Benzene | 0.50 | 5.41 |
| | 0.100 | 5.46 |
| | 0.030 | 5.53 |
| Cyclohexane | 0.150 | 5.41 |
| | 0.100 | 5.48 |
| | 0.030 | 5.58 |

Martin and Besnard [71] have noted a change of 1.7 Hz for the $^3J_{H\text{-}C\text{-}S\text{-}P}$ coupling constant in $CH_3SPOCl_2$. De Jeu *et al.* [24] observed very small changes in the four-bond H—H coupling in acetone (Table 6) which they attributed to solvent effects, but they may well be the result of conformational changes. Fields *et al.* [33] in the course of their study of bis(trifluoromethyl)phosphine found the three-bond P—F coupling to be independent of solvent.

## 5. *Summary of Solvent Effects on Three- and Four-Bond Couplings*

Little can be said about the solvent dependence of three- and four-bond coupling constants. Some change. Others do not. When changes do occur they represent an increase in the coupling constant in solvents of increasing polarity, but there are enough exceptions to this generalization to question its validity. No clear evidence for the primacy of a particular interaction mechanism is evident, nor, for that matter, is there any clear evidence that dipole orientation, or other factors present and apparent for one- and two-bond couplings have any bearing on the subject of the solvent dependence of three- and four-bond couplings.

# IV. Summary and Conclusions

The evidence clearly suggests that all coupling constants are intrinsically solvent dependent. One-bond couplings vary by 1–5% or more. Two-bond couplings between hydrogens change by anywhere from 2–80%, although two-bond couplings involving heavier atoms usually only change by 1–10%. Three- and four-bond couplings present rather ambiguous results. Vicinal H–H couplings either show very little solvent dependence or are in many cases solvent invariant. Three-bond couplings involving heavier atoms in so far as they have been studied appear to vary by 2–15%; occasionally more. In most cases (polar solutes, reaction field or hydrogen bonding interactions, normal dipole orientations) one-bond couplings increase, two-bond couplings decrease and three-bond couplings increase in the absolute sense as the polarity or dielectric strength of the solvent increases.

As implied above, the principal interaction mechanism for polar solutes seems to be the reaction field effect. Specific interactions, notably hydrogen bonding, are also common. For non-polar solutes dispersion interactions seem to predominate. None of the investigations reported to date have developed completely satisfactory solutions to the interaction question, but it appears from the most recent studies that all interaction mechanisms are present in all systems. Most authors have simply reported the dominant effect for the particular case with which they were concerned. Particularly intriguing is the indication that dispersion interactions and reaction field effects produce the opposite affect on coupling constants.

The established solvent dependence of spin-spin coupling constants presents many problems and many opportunities. The general magnitude of the observed changes is sufficient to obscure expected correlations, introduce misleading errors in correlations of various kinds, occasionally suggest incorrect structural assignments and otherwise complicate the use of coupling constants for numerous purposes. Our present understanding of the phenomenon (such as it is) provides a possible means for determining signs of coupling constants, allows the separation of solvent magnetic anisotropy effects from other solvent effects, provides an additional probe into the delicate question of molecular interactions, etc. Additional investigation will hopefully clarify our understanding of the phenomenon, delineate the problems it causes and illuminate the opportunities presented.

# V. References

[1] Ronayne, J., Williams, D. H.: Solvent Effects in Proton Magnetic Resonance Spectroscopy. Annual Review of NMR Spectroscopy, Vol. 2, pp. 83–124, New York 1969.

[2] Laszlo, P.: Solvent Effects and Nuclear Magnetic Resonance. Progr. N.M.R. Spectroscopy 3, 231–403 (1968).

[3] Emsley, J. W., Feeney, J., Sutcliffe, L. H.: High Resolution Nuclear Magnetic Resonance Spectroscopy, Vol. 1, Chaps. 3 and 5. New York: Pergamon Press 1965.

[4] Pople, J. A., Santry, D. P.: Molecular Orbital Theory of Nuclear Spin Coupling Constants. Mol. Phys. 8, 1 (1964).

[5] Pople, J. A., Bothner-By, A. A.: Nuclear Spin Coupling between Geminal Hydrogen Atoms. J. Chem. Phys. 42, 1339 (1965).

[6] Onsager, L.: Electric Moments of Molecules in Liquids. J. Am. Chem. Soc. 58, 1486 (1936).

[7] Diehl, P., Freeman, R.: The Influence of Molecular Shape on Solvent Shifts in the Proton Magnetic Resonance Spectra of Polar Solutes. Mol. Phys. 4, 39 (1961).

[8] Buckingham, A. D., Schaefer, T., Schneider, W. G.: Solvent Effects in Nuclear Magnetic Resonance Spectra. J. Chem. Phys. 32, 1227 (1960).

[9] Baur, M., Nicol, M.: Solvent Stark Effect and Spectral Shifts. J. Chem. Phys. 44, 3337 (1966).

[10] Bayliss, N. S.: The Effect of the Electrostatic Polarization of the Solvent on Electronic Absorbtion Spectra. J. Chem. Phys. 18, 292 (1950).

[11] McRae, E. G.: Theory of Solvent Effects on Molecular Electronic Spectra. J. Phys. Chem. 61, 562 (1957).

[12] Raynes, W. T.: An Empirical Correlation Concerning the Solvent Dependence of Nuclear Spin-Spin Couplings. Mol. Phys. 15, 435 (1968).

[13] Smith, S. L., Cox, R. H.: Solvent Effects on Geminal H–H Couplings: A New Method for Determining Signs of Coupling Constants. J. Chem. Phys. 45, 2848 (1966).

[14] Bell, C. L., Danyluk, S. S.: Solvent Dependence of $^{13}C$–H and $^{13}C$–F Coupling Constants. J. Am. Chem. Soc. 88, 2344 (1966).

[15] Raynes, W. T., Sutherly, T. A.: Linear Electric Field Dependence of $^{13}C$–H Spin-Spin Couplings. Mol. Phys. 17, 547 (1969).

[16] Hammel, J. C., Smith, J. A. S., Wilkins, E. J.: Intramolecular Electric Field Effects in Some Complexes of Acetylacetone. J. Chem. Soc. A, 1969, 1461.

[17] Gil, V. M. S., Teixeira-Dias, J. C. C.: Calculations of Substituent Effects on Directly Bonded $^{13}C$–H Coupling Constants. Mol. Phys. 15, 47 (1968).

[18] Evans, D. F.: Solvent Shifts of Nuclear Spin Coupling Constants due to Hydrogen Bonding. J. Chem. Soc. 1963, 5575.

S. L. Smith

19) Laszlo, P.: Constantes de Couplage et Structure en Resonance magnetique Nucleaire. IV. -Utilisation Practique, en Serie Cyclique et Heterocyclique, des Couplages $^{13}C-H$. Bull. Soc. Chim. France $1966$, 558.

20) Douglas, A. W., Dietz, D.: $^{13}C-H$ Coupling Constants. III. $^{13}C-H$ Coupling in the Vapor Phase and its Dependence on Medium Effects. J. Chem. Phys. $46$, 1214 (1967).

21) Watts, V. S., Goldstein, J. H.: Solvent Effects on $^{13}C-H$ Coupling Parameters and Chemical Shifts of Some Halomethanes. J. Phys. Chem. $70$, 3887 (1966).

22) Martin, G., Castro, B., Martin, M.: Chemie Physique.-Etude par Resonance Magnetique Nucleaire de Quelques Composes Chlores α Functionnels. Compt. Rend. $261$, 395 (1965).

23) Cox, R. H., Smith, S. L.: Solvent Dependent Coupling Constants in Fluoromethanes. J. Mag. Res. $1$, 432 (1969).

24) DeJeu, W. H., Gaur, H. A., Smidt, J.: Solvent Dependence of $^{13}C-H$ and $H-H$ Coupling in Acetone and DMSO. Rec. Trav. Chim. $84$, 1621 (1965).

25) Watts, V. S., Loemker, J., Goldstein, J. H.: Solvent and Concentration Effects on $^{13}C-H$ Coupling Constants and Chemical Shifts in some Dihaloethylenes. J. Mol. Spectry $17$, 348 (1965).

26) Bell, C. L., Danyluk, S. S.: Solvent Dependence of the Chemical Shifts and Coupling Constants of 1,2-dichlorofluoroethylenes. J. Mol. Spectry. $35$, 376 (1970).

27) Rahkamaa, E., Jokisaar, J.: Temperature and Solvent Dependence of $^{13}C-H$ Spin-Spin Coupling Constant $J_{CH}$ in Ethyl Formate. Z. Naturforsch. $A23$, 2094 (1968).

28) Dhingra, M. M., Govil, G., Khetrapal, C. L.: Substitution and Solvent Effects on $^{13}C-^{19}F$ Coupling Constants. Proc. Ind. Acad. Sci. Sect. A, $64$, 91 (1966).

29) Coyle, T. D., Johannesen, R. B., Brinckman, F. E., Farrar, T. C.: Nuclear Magnetic Resonance Studies of Inorganic Fluorides. II. Solvent Effects on $J_{29_{Si}-19_F}$ in Silicontetrafluoride. J. Phys. Chem. $70$, 1682 (1966).

30) Hutton, H. M., Bock, E., Schaefer, T.: Solvent Dependence of $^{29}Si-^{19}F$ Coupling Constant in Silicontetrafluoride. The Importance of Intermolecular Dispersion Interactions. Can. J. Chem. $44$, 2772 (1966).

31) Laszlo, P., Speert, A.: NMR Coupling Constants as Probes into London-Van der Waals Interactions. J. Mag. Res. $1$, 291 (1969).

32) Ebsworth, E. A. V., Sheldrick, G. M.: NMR Spectra of Phosphine, Arsine and Stilbine. Trans. Faraday Soc. $63$, 1071 (1967).

33) Fields, R., Green, M., Jones, A.: Bis(trifluoromethyl)phosphine: Solvent Effects on Nuclear Magnetic Resonance Parameters. J. Chem. Soc. A, $1969$, 2740.

34) Kleiman, Y. L., Morkovin, N. V., Ionin, B. I.: Effect of Solvents and Temperature on Spin-Spin Interaction Constant $^1J_{HP}$ in Dimethyl Hydrogen Phosphite. J. Gen. Chem. R. $37$, 2661 (1967).

35) Vinogradov, L. I., Samitov, Yu. Yu., Yarkova, E. G., Muratova, A. A.: Solvent Effect on Spin-Spin Interaction Constants in Proton Magnetic Resonance Spectra of Some Organophosphorus Compounds Containing P=O (Phosphinylidine) Groups. II. $^1J_{P-H}$ Constants, s-Nature of the P−H Bond and the Value PH. Opt. Spektr. $26$, 959 (1969); Chem. Abstr. $71$, 75583x (1969).

36) Raynes, W. T., Sutherly, T. A., Buttery, H. J., Fenton, C. M.: Solvent Dependence of Nuclear Spin-Spin Coupling Constants. Mol. Phys. $14$, 599 (1968).

37) Paolillo, L., Becker, E. D.: The Effect of Solvent Interactions and Hydrogen Bonding on $^{15}N$ Chemical Shifts and $^{15}N-H$ Coupling Constants. J. Mag. Res. $2$, 168 (1969).

38) Kuhlmann, K., Grant, D. M.: Spin-spin Coupling in the Tetrafluoroborate Ion. J. Chem. Phys. $68$, 3208 (1964).

39) Haque, R., Reeves, L. W.: Coupling Constant and Chemical Shift of Tetrafluoroborate Ion in Mixed Solvents. J. Chem. Phys. $70$, 2753 (1966).

40) Gillespie, R. J., Hartman, J. S.: Change of Sign of the Boron-Fluorine Spin-spin Coupling Constant in the Tetrafluoroborate Anion. J. Chem. Phys. 45, 2712 (1966).

41) Gillespie, R. J., Hartman, J. S., Parekh, M.: Solvent Effects on the Boron-Fluorine Coupling Constant and on Fluorine Exchange in Tetrafluoroborate Anion. Can. J. Chem. 46, 1601 (1968).

42) Dean, P. A. W., Evans, D. F.: Spectroscopic Studies of Inorganic Fluoro-complexes. Part I. The $^{19}$F Nuclear Magnetic Resonance and Vibrational Spectra of Hexafluorometallates of Groups IVA and IVB. J. Chem. Soc. A, 1967, 698.

43) Hutton, H. M., Schaefer, T.: Proton Coupling Constants in Substituted Cyclopropanes. Can. J. Chem. 41, 684 (1963).

44) Shapiro, B. L., Ebersole, S. J., Karabatsos, G. J., Vane, F. M., Manatt, S. L.: Geminal Proton-Proton Coupling Constants in $CH_2$=N-Systems. J. Am. Chem. Soc. 85, 4041 (1963).

45) Shapiro, B. L., Kopchik, R. M., Ebersole, S. J.: Proton NMR Studies of CHDO and $CH_2O$. J. Chem. Phys. 39, 3154 (1963).

46) Shapiro, B. L., Kopchik, R. M., Ebersole, S. J.: Proton Magnetic Resonance Studies of Formaldoxime and its Methyl Ether. J. Mol. Spectry. 11, 326 (1963).

47) Watts, V. S., Reddy, G. S., Goldstein, J. H.: The Variation of the H—H Coupling Constant and Chemical Shifts in α-Chloroacrylonitrile with Concentration and Solvent. J. Mol. Spectry. 11, 325 (1963).

48) Watts, V. S., Goldstein, J. H.: Dependence of Some Ethylenic $J_{gem}$ Values on Solvent and Concentration. J. Phys. Chem. 42, 228 (1965).

49) McLauchlan, K. A., Reeves, L. W., Schaefer, T.: The Sign, Temperature and Solvent Dependence of the Proton Coupling Constant and Chemical Shift in α-Chloroacrylonitrile. Can. J. Chem. 44, 1473 (1966).

50) Schmidt, R. L., Butler, R. S., Goldstein, J. H.: The Role of Polar Factors in Collision Complex Models for the Solvent and Concentration Dependence of Nuclear Magnetic Resonance Parameters. J. Phys. Chem. 73, 1117 (1969).

51) Martin, G. J., Martin, M. L.: Etude Par Resonance Magnetique Nucleaire et Absorbtion Infrarouge de Bromures Vinyliques Purs et en Solution. J. Chim. Phys. 61, 1222 (1964)

52) Martin, M. L., Martin, G. J.: Solvent Effects on the Spin-Spin Coupling Constants between Allenic Protons. J. Mol. Spectry. 34, 53 (1970).

53) Bates, P., Cawley, S., Danyluk, S. S.: Solvent Dependence of Proton-Proton Coupling Constants in Substituted Vinyl Silanes. J. Chem. Phys. 40, 2415 (1964).

54) Blears, D. J., Cawley, S., Danyluk, S. S.: Proton Magnetic Resonance Spectra of Vinyl-metallic Compounds. Solvent Effects upon the Spectra of alpha-chlorovinyl Trichlorosilane and alpha-chlorovinyl Trimethylsilane. J. Mol. Spectry. 26, 524 (1968).

55) Smith, S. L., Cox, R. H.: Solvent Dependence of Proton-Proton Coupling Constants in Styrene Oxide. J. Mol. Spectry. 16, 216 (1965).

56) Cox, R. H., Smith, S. L.: Solvent Dependence of H—H Coupling Constants in 2,2-Dichlorocyclopropylbenzene. J. Mol. Spectry. 21, 232 (1966).

57) Smith, S. L., Cox, R. H.: Solvent Dependent Couplings in Hexachlorobicyclo [2.2.1.] heptenes. J. Phys. Chem. 72, 198 (1968).

58) Finegold, H.: Nuclear Magnetic Resonance Studies of Weak Intermolecular Forces — Medium Effects in Saturated Heterocyclic Rings. J. Phys. Chem. 72, 3244 (1968).

59) Rattet, L. S., Williams, A. D., Goldstein, J. H.: Medium Dependent Geminal Coupling in Open Chain $sp^3$ Hybridized Systems (Acetals). J. Mol. Spectry. 26, 281 (1968).

60) Rattet, L. S., Williams, A. D., Goldstein, J. H.: Dependence of $sp^3$ Geminal Coupling Constants in Acetal and Some Haloacetals on Solvent and Concentration. J. Phys. Chem. 72, 2954 (1968).

185

S. L. Smith

61) Smith, S. L., Ihrig, A. M.: Solvent Effects on H−F Couplings: Dipole Orientation Requirements for Solvent Dependence of Coupling Constants. J. Chem. Phys. *46*, 1181 (1967).

62) Ihrig, A. M., Smith, S. L.: The Solvent and Temperature Dependence of H−H, H−F and F−F Coupling Constants in Difluoroethylenes. J. Am. Chem. Soc. *23* (in press).

63) Frankiss, S. G.: Nuclear Magnetic Resonance Spectra of Some Substituted Methanes. J. Phys. Chem. *67*, 752 (1963).

64) Gutowsky, H. S., Jonas, J., Fu-Mimg, Chen, Meinzer, R.: Intermolecular Interactions and Electron Coupling of Nuclear Spins. J. Chem. Phys. *42*, 2625 (1965).

65) Ihrig, A. M.: The Solvent Dependence of H−H, H−F and F−F Coupling Constants in both Saturated and Unsaturated Systems. Ph. D. Dissertation, Margaret I. King Library, University of Kentucky, Lexington, Kentucky, USA, 1968.

66) Ng, S., Tang, J., Sederholm, C. H.: Effect of Solvent on $^{19}$F Spin-Spin Coupling Constants. J. Chem. Phys. *42*, 79 (1965).

67) McDonald, C. J., Schaefer, T.: Medium Dependence of the Proton Chemical Shift and H−H, H−F and F−F Coupling Constants in 1,1-Difluoroethylene. Can. J. Chem. *45*, 3157 (1967).

68) Alexakos, L. G., Cromwell, C. D.: Nmr Spectra of $ClF_3$ and ClF: Gaseous Spectra and Gas-to-liquid Shifts. J. Chem. Phys. *41*, 2098 (1964).

69) Hatton, J. V., Schneider, W. G., Siebrand, W.: Nuclear Spin-Spin Coupling Involving Heavy Nuclei. The Coupling between Hg$^{199}$ and H$^1$ Nuclei in $CH_3HgX$ and $CH_3CH_2HgX$ Compounds. J. Chem. Phys. *39*, 1330 (1963).

70) Gordon, M., Griffin, C. E.: Solvent Dependence of Geminal Phosphorus-Proton Coupling Constants in Benzylphosphonium Salts. J. Chem. Phys. *41*, 2570 (1964).

71) Martin, G., Besnard, A.: Chemie Physique. − Etude par Resonance Magnetique Nucleaire de L'effet de la Dilution, dans des Solvants nons Polaires, de Composes Organophosphores: Variations de L'effet de Couplage $J_H \ldots P$. Compt. Rend. *257*, 898 (1963).

72) Alger, T. D., Gutowsky, H. S.: Solvent and Temperature Effects on NMR Spectral Parameters in 2-Fluoropyridine. J. Chem. Phys. *48*, 4625 (1968).

73) Crepaux, D., Lehn, J. M.: Nuclear Spin-Spin Interactions Part IX. Relative Signs of the Geminal Nitrogen-Hydrogen Coupling Constants for Doubly Bonded Nitrogen. Mol. Phys. *14*, 547 (1968).

74) Crepaux, D., Lehn, J. M., Dean, R. R.: Nuclear Spin-Spin Interactions. X. Signs of Geminal and Vicinal Nitrogen-Proton Coupling Constants. Stereochemistry and Medium Effects on NH Couplings. Mol. Phys. *16*, 225 (1969).

75) Laszlo, P., Bos, H. J. T.: Effet de Solvents sur les Constantes de Couplage entre Protons Ethyleniques Vicinaux. Tetrahedron Letters *1965*, 1325.

76) Hutton, H. M., Schaefer, T.: Solvent Dependence of Chemical Shift and Coupling Constants in *p*-nitroanisole. Can. J. Chem. *43*, 3116 (1965).

77) Smith, S. L., Ihrig, A. M.: Solvent Dependence of H−H Couplings in Aromatic Compounds. J. Mol. Spectry. *22*, 241 (1967).

78) Erickson, L. E.: Proton Magnetic Resonance Spectra of Substituted Succinic Anhydrides, Acids, Salts and Esters. J. Am. Chem. Soc. *87*, 1867 (1965).

79) Hutton, H. M., Schaefer, T.: Solvent Effects on the Proton Chemical Shift and H−H, H−F Coupling Constants in 1-Chloro-1-fluoroethylene. Reaction Field and Dispersion Interactions. Can. J. Chem. *45*, 1111 (1967).

80) Kumar, A.: NMR Spectra of 1,3-difluoro-4,6-dinitrobenzene and Solvent Effects on Couplings Constants. Mol. Phys. *12*, 593 (1967).

81) Hutton, H. M., Richardson, B., Schaefer, T.: Solvent and Substituent Effects on the H−F Coupling Constants of Some Substituted Fluorobenzenes. Can. J. Chem. *45*, 1795 (1967).

[82] Cooper, M. A.: NMR Spectra of Some Iodo- and Bromofluorobenzenes. Novel Solvent Effect on Ortho Fluorine-Fluorine Couplings. Org. Mag. Res. *1969*, 363.

[83] Rader, C. P.: Hydroxyl Proton Magnetic Resonance Study of Aliphatic Alcohols. J. Am. Chem. Soc. *91*, 3248 (1969).

Received March 15, 1971

# Fortschritte der chemischen Forschung
# Topics in Current Chemistry

## Neuere Bände

**Band 17**
W. Demtröder:
Laser Spectroscopy
With 16 fig. III,95 pages
1971. DM 28,–

**Band 18**
R. C. Bingham/
P. v. R. Schleyer:
Chemistry of Adamantanes
With 4 fig. III,102 pages
1971. DM 36,–

**Band 19**
L. Maier and G. Zon/
K. Mislow: The Chemistry
of Organophosphorus
Compounds I
With 11 fig.,III,94 pages
1971. DM 34,–

**Band 20**
H.J. Bestmann/
R. Zimmermann:
The Chemistry of Organo-
phosphorus Compounds II
With III,147 pages
(In German)
1971. DM 58,–

**Band 21**
L. Eberson/H. Schäfer:
Organic Electrochemistry
With 10 fig. III,182 pages
1971. DM 58,–

**Band 22**
W. Kutzelnigg/G. Del Re/
G. Berthier:
$\sigma$ and $\pi$ Electrons
in Organic Compounds
With 8 fig. III,122 pages
1971. DM 48,–

**Band 23**
M.J.S. Dewar and
W.B. England/L.S. Salmon/
K. Ruedenberg:
Molecular Orbitals
With 40 fig. and 5 tables
III,123 pages
1971. DM 32,–

**Band 24**
H. Fischer and J.F.Labarre/
F. Crasnier: Electronic
Structure of Organic
Compounds
With 12 fig. III,54 pages
1971. DM 18,–

**Band 25**
J. Manassen, R.L. Banks,
W. Strohmeier,
G.-M.Schwab, F.Steinbach:
Catalysis
With 26 fig. III, 154 pages
1972. DM 48, –

**Band 26**
J.L. Margrave/K.G. Sharp/
P.W. Wilson, A. Meller and
G.D. Christian:
Inorganic and Analytical
Chemistry
With 6 fig. III, 112 pages
1972. DM 36, –

**Springer-Verlag**
**Berlin**
**Heidelberg**
**New York**
München · London
Paris · Tokyo · Sydney

# Technische Chemie
# Verfahrenstechnik

**H. Titze:**
**Elemente des**
**Apparatebaues**
Grundlagen. Bau-
elemente. Apparate.
2. Aufl. Mit 241 Abb. XVI,
320 Seiten. 1967
Gebunden DM 49,50
ca. US $ 14.40

**H. H. Buchter:**
**Apparate und**
**Armaturen der**
**Chemischen Hoch-**
**drucktechnik**
Konstruktion, Berech-
nung und Herstellung
Mit 313 Abb. XXIV,
672 Seiten. 1967
Gebunden DM 108,—
ca. US $ 31.20

**H. Kölbel und**
**J. Schulze:**
**Der Absatz in der**
**Chemischen**
**Industrie**
Mit 259 Abb. XV,
732 Seiten. 1970
Gebunden DM 128,—
ca. US $ 37.00

**D'Ans / Lax:**
**Taschenbuch für**
**Chemiker und Physiker**
3. Aufl. In 3 Bänden

1. Band:
**Makroskopische**
**physikalisch-chemi-**
**sche Eigenschaften**
Hrsg. von E. Lax unter
Mitarbeit von
C. Synowietz
XVI, 1522 Seiten. 1967
Gebunden DM 68,—
ca. US $ 19.60

2. Band:
**Organische**
**Verbindungen**
Hrsg. von E. Lax unter
Mitarbeit von
C. Synowietz
VIII, 1177 Seiten. 1964
Gebunden DM 48,—
ca. US $ 13.90

3. Band:
**Eigenschaften von**
**Atomen und**
**Molekeln**
Hrsg. von K. Schäfer und
C. Synowietz.

Mit 112 Abb. VIII,
670 Seiten. 1970
Gebunden DM 48,—
ca. US $ 13.90

**H. Röpke und**
**J. Riemann:**
**Analogcomputer in**
**Chemie und Biologie**
Eine Einführung
Mit 198 Abb. VII,
184 Seiten. 1969
Gebunden DM 38,—
ca. US $ 11.00

■ **Bitte fordern Sie**
**Prospekte an!**

**Springer-Verlag**
**Berlin**
**Heidelberg**
**New York**
München · London · Paris
Tokyo · Sydney

In kritischen Übersichten werden in dieser Reihe Stand und Entwicklung aktueller chemischer Forschungsgebiete beschrieben. Sie wendet sich an alle Chemiker in Forschung und Industrie, die am Fortschritt ihrer Wissenschaft teilhaben wollen.

In der Regel werden nur Beiträge veröffentlicht, die ausdrücklich angefordert worden sind. Schriftleitung und Herausgeber sind aber für ergänzende Anregungen und Hinweise jederzeit dankbar. Manuskripte können in den „Fortschritten der chemichen Forschung" in Deutsch oder Englisch veröffentlicht werden.

Jeder Band der Reihe ist einzeln käuflich.

This series presents critical reviews of the present position and future trends in modern chemical research. It is addressed to all research and industrial chemists who wish to keep abreast of advances in their subject.

As a rule, contributions are specially commissioned. The editors and publishers will, however, always be pleased to receive suggestions and supplementary information. Papers are accepted for "Topics in Current Chemistry" in either German or English.

Any volume of the series may be purchased separately.